Trace Elements and Their Effects on Human Health and Diseases

Edited by Daisy Joseph

Published in London, United Kingdom

IntechOpen

Supporting open minds since 2005

Trace Elements and Their Effects on Human Health and Diseases
http://dx.doi.org/10.5772/intechopen.91798
Edited by Daisy Joseph

Contributors
Naheen Waseem Abdul Sadique Sheikh, Tusharbindu R. Desai, Satish B. Kosalge, Alok S. Tripathi, Deepak S. Mohale, Anil P. Dewani, Olumayede Gbenga Emmanuel, Babalola Bolanle, Itebimien Oghenovo, Harpreet Singh Kainth, Deeksha Khandelwal, Ranjit Singh, Sanjiv Puri, Gurjeet Singh, Zohra Lamari, Houria Negache, Wojciech Roczniak, Magdalena Babuśka - Roczniak, Elżbieta Cipora, Barbara Brodziak - Dopierała, Biju Prava Sahariah, Jyoti Kant Choudhari, Jyotsna Choubey, Mukesh Kumar Verma, Anandkumar Jayapal, Daisy Joseph

First published in London, United Kingdom, 2021 by IntechOpen
IntechOpen is the global imprint of INTECHOPEN LIMITED, registered in England and Wales, registration number: 11086078, 5 Princes Gate Court, London, SW7 2QJ, United Kingdom
Printed in Croatia

British Library Cataloguing-in-Publication Data
A catalogue record for this book is available from the British Library

Additional hard and PDF copies can be obtained from orders@intechopen.com

Trace Elements and Their Effects on Human Health and Diseases
Edited by Daisy Joseph
p. cm.
Print ISBN 978-1-83968-644-3
Online ISBN 978-1-83968-645-0
eBook (PDF) ISBN 978-1-83968-646-7

For EU product safety concerns:
IN TECH d.o.o., Prolaz Marije Krucifikse Kozulić 3, 51000 Rijeka, Croatia, info@intechopen.com or visit our website at intechopen.com.

Meet the editor

Dr. Daisy Joseph is a scientist at the Bhabha Atomic Research Centre (BARC), Mumbai. She obtained a Ph.D. in Physics from Mumbai University with her thesis, "Study of X-ray spectra by photo, proton and heavy ion excitation." She is an expert in energy dispersive X-ray fluorescence (EDXRF) and particle-induced X-ray emission (PIXE) and has around 100 publications on these topics in peer-reviewed journals. She has also written a book chapter and edited the book *Synchrotron Radiation.* She is the associate editor for *Advances in Basic Science* and *BioScience.* She has guided several Ph.D. students from India. She has represented BARC, India, at many international conferences and workshops in her field. She is a collaborator with institutes in India, including Christ University, Bangalore; Karnatak University, Dharwad; Punjab University, Chandigarh; SR & BGNR College of Science and Arts, Khammam, Andhra Pradesh; National Institute of Manipur, Langol; Manipur University Department of Physics; Mumbai University; and Birla Institute of Technology & Science (BITS), Pilani. She is a reviewer for several journals including *X-ray Spectrometry, Journal of Radiation Isotopes, Bio-Science Journal, and Spectroscopy Letters.*

Contents

Preface

Trace elements have always been an interesting area of research, as the human body contains a large number of trace elements that fully control enzymes and hormone production. This book examines the useful as well as harmful effects of trace elements on both humans and the environment. It includes seven chapters divided into two sections. Section 1, "Trace Elements in Biology," includes chapters on the effects of trace elements on breast cancer, respiratory diseases, tissue impairment, and iron-related diseases. Section 2, "Trace Elements in Environmental Problems" includes chapters that focus on environmental issues, such as urban particulate matter and trace elements in medicinal plants.

This book is a useful resource for students and researchers involved with human health and diseases. I hope this book provides useful information for carrying out novel research in this field.

Dr. Daisy Joseph
Nuclear Physics Division,
Bhabha Atomic Research Centre,
Mumbai, India

Section 1

Trace Elements in Biology

Chapter 1

Introductory Chapter: Tracelement Effects in Biological Applications Using Photon Induced (EDXRF), Proton Induced (PIXE) and Synchrotron Induced (EXAFS) X-Ray Spectrometry

Daisy Joseph

1. Introduction

Heavy metals are dangerous to human health. Heavy metals are Sb, As, Bi, Cd, Ce, Cr, Co, Cu, Ga. Trace elements occur in natural environments in small amounts and when present in sufficient concentrations, are toxic to living organisms [1]. Trace elements enter the ecosystems via direct discharges from industrialization processes, sewage sludge, atmospheric deposits and agricultural practices including application of pesticides or fertilizers [2–4]. They can be transferred from sediments to benthic organisms and then become a potential risk to human consumers through the food chain [5]. We have a X-ray emission spectrometer. Energy Dispersive X-ray Fluorescence and Proton Induced X-ray Emission, in which trace elements (micronutrients as well as toxic elements) were anlysed.

2. Application of EDXRF and PIXE in biosciences

The large number of applications of XRF in our laboratory has been carried out in Biosciences. They can be summarized as follows.

2.1 Detection of trace elements in Indian spices

Concentrations of K, Ca, Mn, Fe, Cu, Zn, Rb and Sr were determined in Indian Spices namely pepper, clove, cardomon, cinnamon, and cumin using Cd ^{109} radioisotope source induced XRF. The levels of K and Ca were highest in clove and cinnamon. Rubidium and Strontium were found in all spices except cinnamon. Chromium and titanium were found only in pepper [6].

2.2 Determination of mercury and arsenic in Indian ayurvedic medicines using EDXRF

Elemental concentration in some herbal medical products, produced by different *ayurvedic* pharmacies in India, was determined using Energy Dispersive X-Ray

IntechOpen

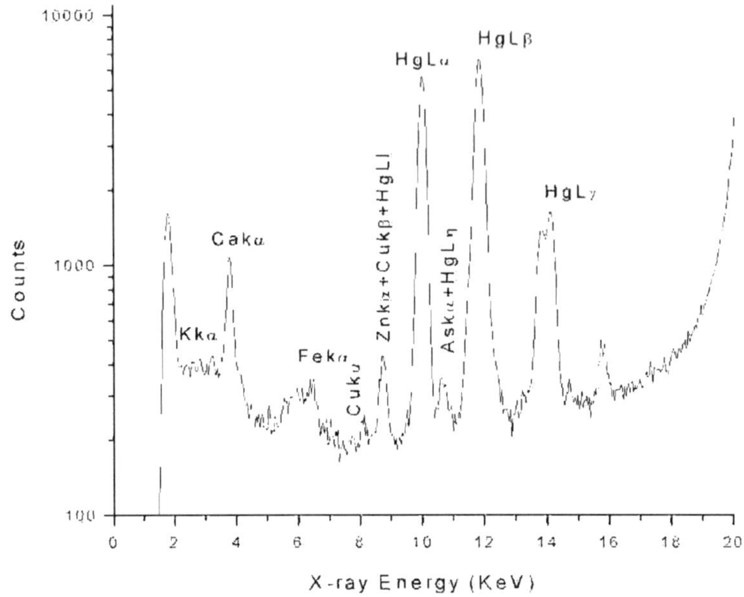

Figure 1.
A typical EDXRF spectrum of an Indian ayurvedic medicine Balguti Kesaria.

Fluorescence Spectrometry (EDXRF). It is mandatory to look into the effects of these heavy metals being administered in the body functions before taking them over a long period of time (**Figure 1**) [7].

2.3 Drought tolerant and susceptible genotypes of sorghum plants

Drought tolerant and susceptible genotypes of sorghum plants were analyzed by EDXRF technique to study the correlation of trace elements with drought tolerance capacities for sorghum plants. Samples prepared from mature seeds, young seedlings and old plants were analyzed using ^{109}Cd radioisotope source. The elements such as K, Fe, Cu, Zn, Rb and Sr and Y were seen to be present in different quantities in various samples. K and Fe concentrations were found to be more in the tolerant genotype as compared to the susceptible type. Concentration of Fe decreased with maturity in the tolerant group while it increased with maturity in the susceptible group. The genotype Arfa Gadamak (AG) showed a distinct abnormality in its young seedling with high level of Zn. In conclusion, the drought tolerant and susceptible genotypes of sorghum genotypes (cultivated in Sudan) exhibited varying levels of trace elements. The drought tolerant genotypes of sorghum seeds exhibited high K and Fe concentrations as compared to susceptible genotypes. In seedlings Fe concentration decreased with maturity in the tolerant group while it increased in the susceptible variety [8].

2.4 PIXE studies of blood Pb levels in children of the Dharavi slum areas in Mumbai

PIXE was used to study lead levels in blood samples of children from Dharavi slum areas. Blood lead levels of children admitted to Sion Hospital, Bombay (India), from the adjoining Dharavi slum areas. Blood samples were collected from 36 children with suspected lead poisoning (indicator was acute anemia) and from 20 control children. The analysis showed that the lead concentration of the patients varied from 0.1 to

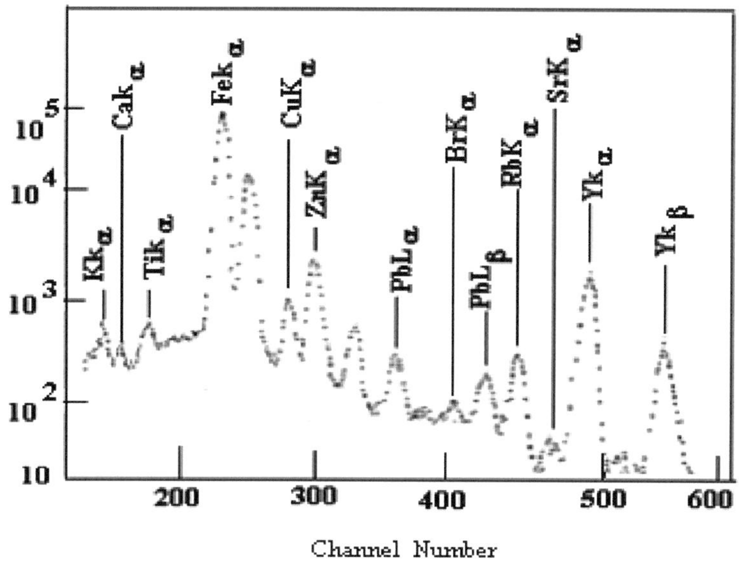

Figure 2.
PIXE spectrum of blood of a lead poisoned child.

6.0 μg ml^{-1}. In addition to lead, K, Ca, Fe, Cu, Zn, Se, Br and Rb were also detected simultaneously, of which the concentrations of Fe, Cu, Zn, Se, Rb and Pb were determined. The high blood lead levels of the children from this area may be ascribed to environmental pollution due to heavy vehicular traffic and industrial sources. Pb was found to high even in normal children due to vehicular exhaust containing lead in petrol. **Figure 2** shows the PIXE spectrum of a Pb poisoned child [9].

2.5 Uranium extraction from cynobacterial cells

Cyanobacterial cell organisms grown in Uranyl nitrate and Uranyl Carbonate were determined for their Uranium uptake by ^{109}Cd induced EDXRF during different time intervals. It was found while elements such as K, Ca, Cr, Mn, Fe and Zn were present in small quantities was seen to be present in significant amounts after 2 hours of uptake and it became saturated in 5 hours after which the uptake reduced and became minimum.

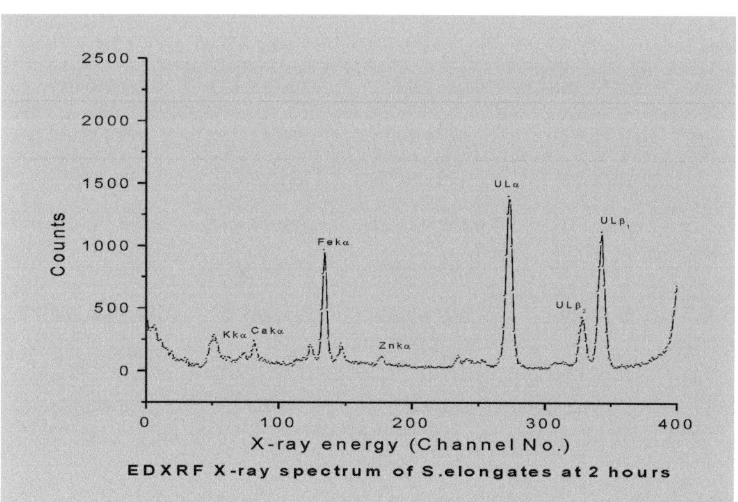

2.6 Tracelement variation in renal failed patients

In a separate experiment of application of PIXE in bio-medical research, blood samples of Patients with Chronic renal failure were analyzed for trace element abnormalities and the results showed marked differences in patients before and after dialyses. PIXE being more sensitive in the low Z region due to its higher cross section is an ideal technique for bio-science applications. The knowledge that small amounts of metals which are needed in the diet goes back several hundred years to the discovery of a requirement for iron. It is well established that some elements such as I, Cu, Mn, Zn, Se and Mb in trace quantity are needed in physiology. Very recently, the six "newer trace elements", tin, vanadium, fluorine, silicon, nickel, and arsenic, were discovered to have nutritional requirement. Recommended Dietary Allowances (RDA) and requirements have been set for iron, iodine, and zinc. For copper the issuance of an RDA can be established. Only tentative recommendations for the "newer" trace elements in the form of a range of values can be presently proposed. To establish these recommendations more firmly, knowledge of the content of each of these trace elements in the diet is necessary. A detailed study of trace element abnormalities in serum of patients was carried out by PIXE in the case of non-dialyzed, hemodialysed and post-transplantation. Forty-two patients and eight healthy controls selected for this study were grouped on the basis of their Serum Creatinine (SC) levels. Serum samples of these subjects were excited by protons from the Van de Graff accelerator of 2.5 MeV energy and the Characteristic X-rays were detected by Si (Li) detector (**Table 1**).

In the case of renal failed patients the exact mechanism of trace element disturbance is not known. Reduced renal excretion increased oral intake and Globin Insulin (G.I.) absorption and contamination as well as loss across the hemodialysis membranes have been incriminated Considerable variation in plasma and tissue concentrations of trace elements have been found in different geolgraphical areas due to variation in water and soil content. Patients on dialysis therapy had elevated serum copper levels. Butamante reported similar results and attributed liberation of copper from the dialysis membrane as a cause for hypercupremia. Successful renal transplantation resorted serum copper to normal levels. A well established observation is that serum brooming levels are not different from normal in non dialyzed chronic renal failure patients, but those on dialysis show subnormal levels. Transference of bromine from blood into the

Elements	I	II	III	IV	V
Ni	.0167 ± .0164	0.034 ± 0.022	0.030 ± 0.033	0.066 ± 0.051	0.042 ± 0.027
Cu	.083 ± 0.16	0.83 ± 0.25	0.97 ± 0.33	1.08 ± 0.37	0.71 ± 0.28
Zn	1.12 ± 0.93	1.06 ± 0.25	1.30 ± 0.01	1.38 ± 0.67	1.13 ± 0.47
Se	.035 ± 0.017	0.032 ± 0.029	0.035 ± 0.011	0.033 ± 0.018	0.030 ± 0.012
Br	3.66± .154	3.383± 0.241	0.599± 0.417	0.122± 0.127	0.281± 0.255
Rb	.178 ± 0.101	0.297 ± 0.177	0.246 ± 0.101	0.172 ± 0.109	0.263 ± 0.114
Sr	.029 ± 0.044	0.132 ± 0.088	0.204 ± 0.089	0.111 ± 0.214	0.088 ± 0.082
Pb	.173 ± 0.143	0.279 ± 0.137	0.347 ± 0.236	0.797 ± 0.537	0.400 ± 0.260

Table 1.
Values in mean (µg/ml) ± S.D. of concentrations of trace elements in various groups.

dialysate could be responsible for this deficiency. Transplantation resulted in rise of bromine towards normal. As a result of loss of renal function, Sr which depends on the kidney for elimination is probably retained accounting for the significantly elevated serum concentration of Sr in patients of CRF. Patients on hemodialysis and those who received a successful renal allograft had SR concentration within the normal range. Chronic Renal insufficiency did not result in accumulation of lead in our study similar to that observed by Thomson et al. Contamination of the dialysate or dialysate delivery system possibly resulted in elevation of lead concentration while on dialysis. Following renal transplantation the serum lead levels went down as compared to lead levels while on dialysis, but were still significantly elevated when compared to normal. Mobilization of lead which was sequestered in the tissues while the patient was on dialysis into the serum, to be excreted via the kidneys could explain the high lead levels. It could also be possible that the transplanted kidney has not yet attained a normal function with regard to lead excretion and with passage of time normal serum lead level would be achieved. Low and normal serum zinc levels have been reported in non dialysed patients, but in dialyzed patient elevated subnormal and almost normal concentration have been described. An unrestricted dietary protein intake (45–50 g/day) and normo-proteinemia could explain the lack of hypozincemia studied by Mansouri et al., in patients who were on a protein restricted diet (20–30 g/day) and had significant hypoproteinemia. Our study showed a slight, but non significant elevation of serum zinc in hemodialysed patients, probably as a result of contamination of the dialysate or dialysate delivery system with zinc [10].

2.7 Synchtrotron based EXAFS on Mercury based Indian herbomineral drug.

An Indian herbomineral drug was characterized for its trace elements by radioisotope induced EDXRF.The drug contains minerals like mercury, sulfur and arsenic disulfide, along with herbs By XRF As, Hg, Fe, Ca were detected. S was not detected as S X-rays get absorbed in the detector window. However AsS and HgS was seen in EXAFS spectra. There is no peak in the experimental data corresponding to the 2nd shell of HgS viz.., the Hg-Hg shell (corresponding to the blue lines). It may happen that Hg is not present in the form of HgS in the samples, instead Hg forms bonds with S, C or O present in the herbs. The peaks near 3 Å for samples 4 & 6 and near 1.5 Å of sample 9 might be due to bond formation of Hg with C or O atoms of the herbs. Sample 6 shows maximum disorder and the Hg-S bond length shows an increasing trend from sample 4 to sample 9. A more comprehensive report is underway regarding its structure after a complete analysis of EXAFS.spectra.

The XRF results for the herbomineral samples are as follows:

Sample No	Major analyte	Trace
Sample 4	Hg	Y
Sample 5	Hg and As	Fe
Sample 6	Hg and As	Fe
Sample 7	Hg and As	Fe
Sample 8	Hg and As	Fe and Ca
Sample 9	Hg and As	Fe and Cu

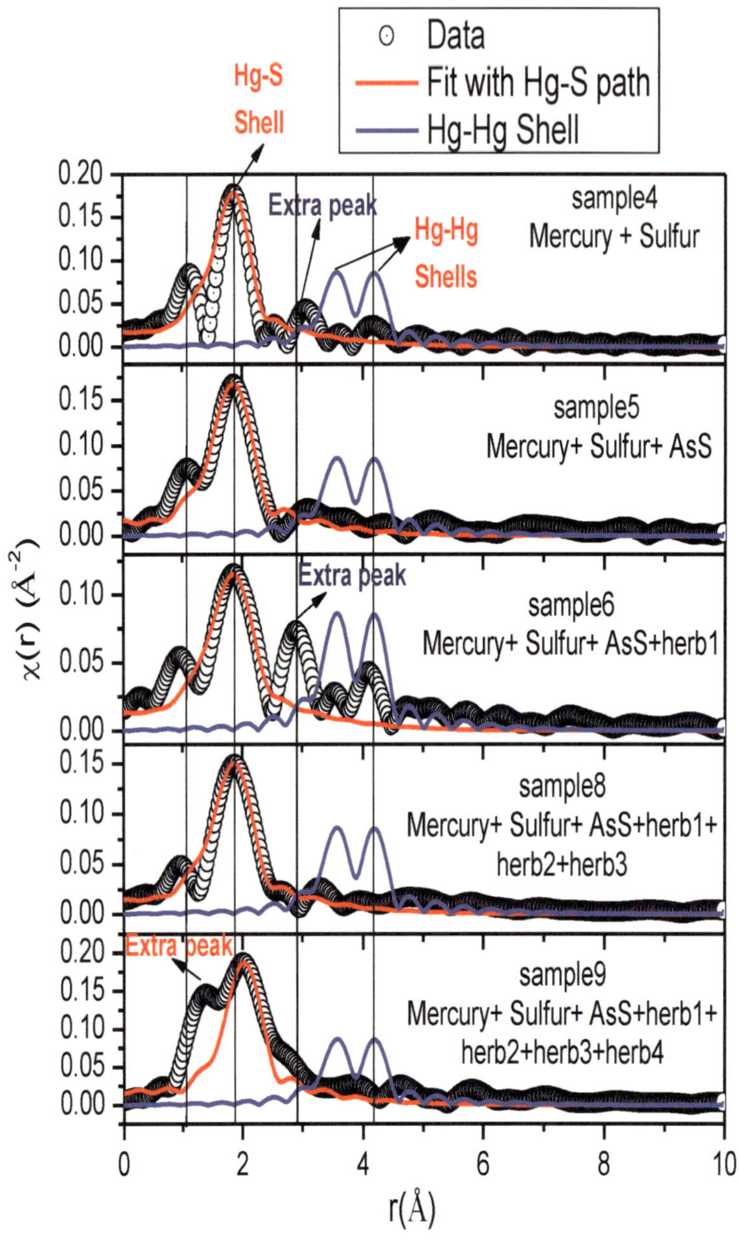

3. Conclusion

Though trace elements are required in minimal quantities their presence in the optimal amount is essential for the normal physiological functioning of the body. They are one of the corner stone's in maintenance of biodynamic of the body. Both, excess and the deficiency states lead to initiation, promotion, and progression to various disease processes. The present paper has thoroughly discussed trace elements, as this area is away from the deserved attention. Thus, a comprehensive understanding of these trace elements is essential and significant for disease control and maintaining optimal health and X-ray Emission Techniques such as EDXRF, PIXE and EXAFS have shown to be good diagnostic tools for determining traceelements in Biological samples.

Author details

Daisy Joseph
Nuclear Physics Division, BARC, Trombay, Mumbai, India

*Address all correspondence to: djoseph@barc.gov.in

References

[1] Adriano DC. Trace elements in terrestrial environments: biogeochemistry, bioavailability, and risks of metals. 2 New York: Springer; 2003.

[2] Han FX, Banin A, Su Y, Monts DL, Plodinec MJ, Kingery WL, Triplett GB. Industrial age anthropogenic inputs of heavy metals into the pedosphere. Naturwissenschaften. 2002;89:497-504. [PubMed]

[3] Han FX, Patterson WD, Xia Y, Sridhar BBM, Su Y. Rapid determination of mercury in plant and soil samples using inductively coupled plasma atomic emission spectroscopy, a comparative study. Water, Air and Soil Pollution. 2006;170:161-171.

[4] Han FX, Su Y, Monts DL, Plodinec MJ, Banin A, Triplett GB. Assessment of global industrial-age anthropogenic arsenic contamination. Naturwissenschaften. 2003;90:395-401. [PubMed]

[5] Soto-Jiménez MF, Arellano-Fiore C, Rocha-Velarde R, Jara-Marini ME, Ruelas-Inzunza J, Páez-Osuna F. Trophic transfer of lead through a model marine four-level food chain: *Tetraselmis suecica*, *Artemia franciscana*, *Litopenaeus vannamei*, and *Haemulon scudderi*. Archives of Environmental Contamination and Toxicology. 2011;61(2):280-291. [PubMed]

[6] D.Joseph, M.Lal, H.N.Bajpai*and P.K.Mathur,Levels of trace elements of a few Indian spices by EDXRF, J.of Food Science and Technology, Vol.36, No.3, 264-265, (1999)

[7] D.Joseph,Determination of Mercury and Arsenic Content in a few Indian Ayurvedic medicines by EDXRF, International Journal of PIXE, Vol. 19, Nos. 1 & 2 (2009) 77–89,Ó World Scientific Publishing Company

[8] D.Joseph, A.H.AbuAssar,P. Suprasanna, A.Saxena,R.K. Choudhury V. Bapat, Study of trace element correlations with drought Tolerance in different Sorghum Genotypes using Energy Dispersive X-ray Fluorescence (EDXRF) technique. Biol. Trace. Element Res. (2002) Vol. 85, No.3, 252-268

[9] Madan Lal, Daisy Joseph, R.K.Choudhury, H.N.Bajpai , Study of blood levels in children by Proton Induced X-ray Emission, Science of Total Environment, 103(1991), 209

[10] Madan Lal, H.N.Bajpai, Daisy Joseph and P.K.Patra Study of Trace Element Abnormalities in Blood of Chronic Renal Failure (CRF) patients by proton induced X-ray emission (PIXE),International Journal of PIXE, Vol 3, No.3, (1993) 229-236.

Chapter 2

Role of Trace Elements in Breast Cancer and Their Characterization Using X-Ray Fluorescence Techniques

Harpreet Singh Kainth, Deeksha Khandelwal, Ranjit Singh, Gurjeet Singh and Sanjiv Puri

Abstract

Breast cancer is the most common serious disease that occurs in the human body. Trace elements have an important function in biological and metabolism processes including activation or inhibition of enzymatic reaction, reactive oxygen species (ROS), competition between trace elements and metal proteins for binding positions and modifications in the permeability of cellular membranes which influence carcinogenic processes. A significant association between the abnormal concentration of trace elements and breast cancer has been found in many studies using XRF techniques like energy dispersive X-ray fluorescence (EDXRF), particle induced X-ray emission (PIXE), total reflection X-ray fluorescence (TXRF), wavelength dispersive X-ray fluorescence (WDXRF) and synchrotron induced X-ray fluorescence (SRIXE). This chapter considers trace elements like Fe, Cu, Zn, Cr, Cl, Ca, P, S, K, Na, Mg, Se, As and Sr. from the standpoint of their role as either inhibitory or causative agents of breast cancer. XRF techniques and sample preparation methods for analysis of biological samples are also reviewed.

Keywords: breast cancer, trace elements and XRF techniques

1. Introduction

In the human body, sometimes cells begin to grow out of control and divide into a large number of abnormal cells usually known as cancer cells. Cancer is a multifactorial complex disease. These cancer cells can spread or metastasize from one part of the body to another part and damage the patient's quality of life. Cancer is of various types like kidney cancer, colon cancer and lung cancer, etc. Among all these, breast cancer is the most common that occurs in the human body system. Generally, this type of cancer is commonly found in women. It is widely accepted that breast cancer is hormonally influenced, with most of the risk factors associated with the exposure of breast to stimulatory effects of female reproductive hormones, mostly estrogens, leading to increased cellular proliferation due to which normal cells gets converted into breast cancerous cells. Only 5–10% of all breast cancer cases are due to genetic factors i.e., the inheritance of mutations in breast cancer susceptible genes (BRCA1 and BRCA2). Rest of the cases are having hormonal and non-hormonal non-genetic risk factors. Non-hormonal cases also indirectly tied to modulation of estrogens exposure.

Breast cancer is the most diagnosed cancer among females worldwide, having recent estimates of 2.1 million new incidence and 630,000 deaths in 2018 [1]. Most incidences of breast cancer are reported in the countries with higher Human Development Index (HDI) alleging westernization of lifestyle linked to menstrual characteristic (age at menarche and menopause and type of menopause), Reproduction factors (older ages at first birth, nulliparity, giving fewer births), exogeneous hormone intake (oral contraceptive pills, Menopausal harmonic therapy), medication (fertility drugs, Diethylstilbesterol), nutrition factors (high fat intake during adolescence, alcohol), anthropometric factors (rapid height growth during childhood and adolescence, high body mass index (BMI), body fat distribution), smoking [2]. Engagement in regular physical activities, to avoid consumption of alcoholic beverages, Breastfeeding with longer duration, balanced diet (fruits and vegetable, soy) is some of the important factors to reduce the risk of breast cancer. Although breast cancer incident rates are highest in economically developed than developing countries, the reverse is true for mortality rates, reflecting limited screening and less effective treatments in such areas. In year 2018, highest incidence rate has been recorded in Australia/New Zealand while highest mortality has been estimated in Melanesia [1].

Trace elements and their role in the cancer process have been a matter of great concern and early reports given by various researchers have proved that there is a relation between trace elements and cancer which play a key role in the biological and metabolic processes in the human body. It is reasonable to assume that the abnormal levels of these trace elements lead to the development of cancer in the human body system. Furthermore, the excess and deficiency of trace elements induce the formation of reactive oxygen species (ROS). It is believed that ROS lead to the formation of almost all types of cancer. Generally, these are divided into two groups: (a) Free oxygen radicals (b) non-radicals. The International Agency for Research on Cancer (IARC) has suggested the list of elements that show the carcinogenetic properties. These elements are Be, Cr, Co, Ni, As, Cd, Sb, Pb, Hg, Pt, Mn, Fe, Cu, Zn, Se and Sr. The abnormal levels of ROS disturb the biological processes and metabolic activities which results in the unchecked normal cells growth into cancerous cells [3]. For the detection of these trace elements, analytical techniques like energy dispersive X-ray fluorescence (EDXRF), total reflection X-ray fluorescence (TXRF), synchrotron induced X-ray fluorescence (SIXRF) and proton induced X-ray fluorescence (PIXE) have been widely used [4–9].

The aim of the present chapter is to discuss the role of trace elements in human breast cancer and the techniques used for quantitative elemental analysis of cancer samples.

2. Methods and materials

X-ray Fluorescence (XRF) is a well-established non-destructive analytical technique for quantitative as well as qualitative determination of elemental composition in samples independent of their physical and chemical forms. In XRF, either electrons or photons (X-rays/γ-rays) used as the excitation source, are incident on the sample thereby exciting the atoms of the elements present in the sample [10, 11]. The intensity of characteristic X-rays and scattered photons resulting from the photon-atom interaction processes are detected and measured using energy dispersive X-ray fluorescence (EDXRF) spectrometers. In EDXRF, the characteristic X-rays are not diffracted spatially and are detected by a detector with signal processing electronics. An entire spectrum can be acquired virtually simultaneously in the EDXRF technique so that the detection of most of the elements across the periodic table can be possible within a few seconds. It is mainly a non-destructive chemical analysis technique and a great variety of non-portable and portable experimental set-ups are available. The triaxial geometry of non-portable spectrometers allows radiations (not monochromatic) of selected energy range, depending upon the secondary target. Si (Li) or Ge detectors, cooled by liquid nitrogen [12, 13], are generally used in these types of set-ups. As far as portable spectrometers are concerned, radioactive sources or X-ray tubes are commonly used. To reduce the noise level, detectors are cooled by the Peltier effect. These types of detectors have limitation of poor sensitivity, so trace elements of biological samples were studied less [14]. The wavelength dispersive X-ray fluorescence (WDXRF) technique is very important for the routine elemental analysis for the quality control of various materials. This technique is based on the Bragg's law, $n\lambda = 2d\sin\theta$, where n is an integer determined by the order of diffraction, θ is the scattering angle, d is the inter-planar distance of crystal lattice and λ is the wavelength of impinging radiation. On being incident on a crystalline sample, they are scattered in a peculiar fashion by the atoms undergoing constructive interference, which occurs only if the electromagnetic radiation or subatomic particle waves have the wavelength comparable to the atomic spacing. In this technique, amplification in the signal is obtained due to constructive interference of detected X-rays which obey the Bragg's condition. The crystal monochromator is one of the key parts in WD spectrometers. WDXRF technique uses optimized analyzer crystals and detectors to separate and count the emitted discrete X-ray wavelengths using diffraction from a crystal with a very high degree of resolution. This technique is more stable for analytical accuracy and precision even in performing the chemical analysis as compared to others. Total reflection X-ray fluorescence (TXRF) [15] and micro X-ray fluorescence (μXRF) [16] are the advanced variants of EDXRF. TXRF utilizes the property of total external reflection. In TXRF a fine, collimated and almost parallel X-ray beam from the X-ray tube falls on a smooth polished surface of the target sample in the form of a thin layer of a few nm thickness, at a grazing angle below the critical angle of the surface and gets totally reflected. Due to this condition, a totally reflected beam reduces scattering and absorption of the incident beam in the photon absorption matrix of a sample. This leads to a largely enhanced peak to background ratio, significantly amended fluorescence yield and consequently much better sensitivities to elements present even in ultra-trace levels. The improved detection limits of TXRF make it a valuable tool for trace and ultra-trace element analysis. The TXRF technique is more sensitive due to the use of glancing angle and destructive to some extent as compared to both the EDXRF and WDXRF techniques. **Figures 1** and **2** show the X-ray emission spectra of normal and abnormal breast tissue/blood samples obtained by non-destructive EDXRF, WDXRF and TXRF techniques. In μXRF, X-rays generated by the X-ray tube are converged at a small region ~10 μm on the sample surface by polycapillary lens (an X-ray focusing system) that exploits the phenomenon of multiple total external reflection in array of small hollow glass tubes [17]. The polycapillary lens increases the intensity and spatial resolution of X-rays irradiating on the sample. Also, irradiating the sample with X-rays micro focused only on the target position enhances the signal-to-background ratio by reducing fluorescence X-rays generated from adjacent areas. The photon microprobe is the best technique of the future for material information because of the very low deposit in the matter and its variety of interactions. One can use this technique in various element mapping applications of X-ray fluorescence.

Synchrotron induced X-ray fluorescence (SIXRF) offers distinct advantages over other XRF techniques as the synchrotron radiation has been used as a

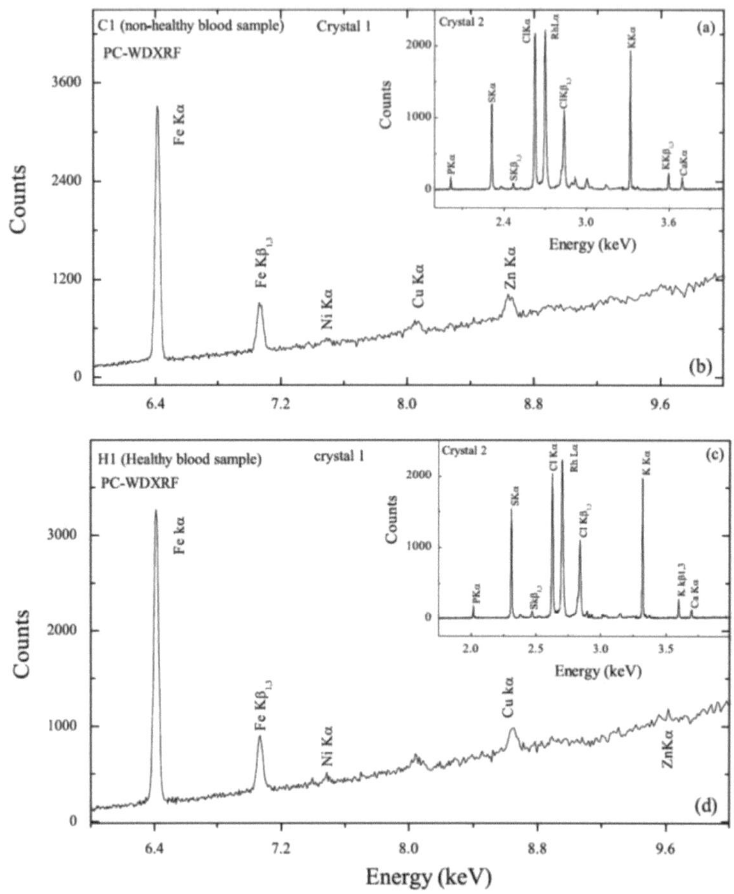

Figure 1.
Insert (a) and (c) show the spectrum of non-healthy (C1) and healthy (H1) blood samples of patients taken by crystal 2. Caption (b) and (d) represent spectrum of non-healthy (C1) and healthy (H1) blood samples of patients taken by crystal 1 of PC-WDXRF spectrometer [32].

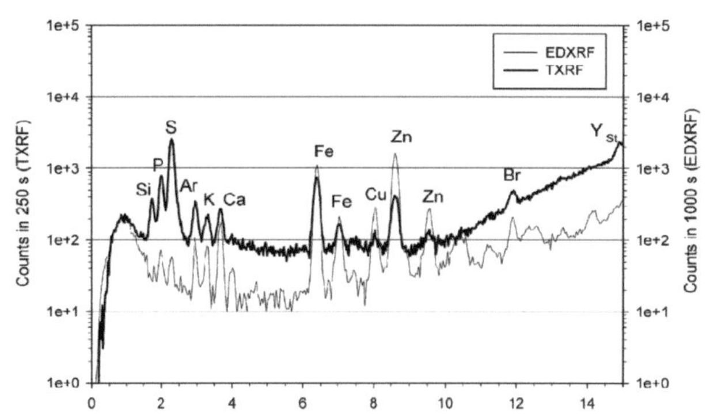

Figure 2.
X-ray emission spectra of normal and abnormal breast tissue/blood samples obtained by non-destructive EDXRF and TXRF techniques [31].

powerful X-ray source [18]. The synchrotron radiation source is characterized by a high degree of polarization and pulse height, high collimation, low emittance, reliability in energy by monochromatized emission. It has become an important tool in various fields of research. The synchrotron source provides the combination of high flux and low divergence which is crucial for the massive success of experiments in the field of SIXRF. An important property of the synchrotron radiation is linear polarization which allows the SIXRF setup to detect the concentrations of specific elements present at trace levels due to a significant reduction in background level produced from Compton scattering. Almost complete repression of Compton scattering can be achieved by locating the detector at 90° with respect to the synchrotron beam in the plane of polarization. Hence the improved detection limit can be obtained using synchrotron radiation. The directionality and brightness of synchrotron radiation provide a superlative capability for micro-beam analysis. **Figure 3** represents the X-ray emission spectra of normal and cancerous blood serum by non-destructive synchrotron based XRF technique.

Particle induced X-ray emission (PIXE) is well known technique for the elemental analysis and high cross-sections of the elements. In recent years, most of the scientists use this technique for the biological samples. Due to the low level of continuum background, it produced better results than other techniques. PIXE opens up a new era in the field of biological samples where measurements of low Z elements are possible at microscale level. It is also well known that the microbeam PIXE is the technique that offers results in ppm level with high sensitivity and the

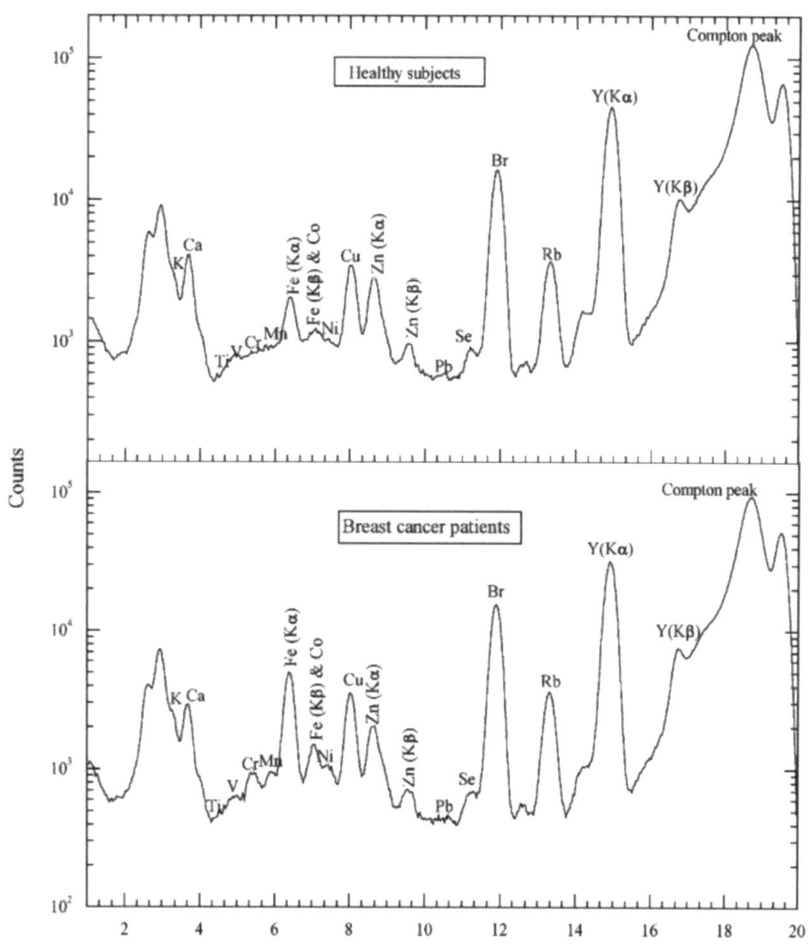

Figure 3.
X-ray emission spectra of normal and cancerous blood serum by non-destructive synchrotron based XRF technique [38].

size of the beam is smaller than the biological sample cell dimensions. In PIXE, the active protons (MeV) excite the target atoms to produce the X-ray spectrum by the inner shell decay process producing X-ray. The emitted X-ray energies are the characteristics of the elements and are proportional to the mass of that element present in the sample for further analysis [19–21]. **Figure 4** illustrates the normal and abnormal breast cancerous tissue obtained from PIXE spectrum. X-Ray absorption spectroscopy is a technique in which a core electron is excited to an empty state of LUMO and continuum, known as X-Ray Absorption Near Edge Spectroscopy (XANES) and extended X-ray absorption fine structure (EXAFS), respectively [22]. Mostly synchrotron is used as X-Ray source, but laboratory based commercial system are also available [23]. XANES spectra reveals average oxidation state, local coordination environment, chemical speciation and symmetry of the metal site while the EXAFS delineates the identity, number, distance of neighboring or adjacent atoms from the excited atom [24, 25]. XAS techniques acquire ascendency over X-Ray crystallographic techniques in the sense that local structural information around an element can be unraveled even from disordered samples such as powders and solutions. XANES spectroscopy has been used to examine the oxidation state of Zn, Fe and Cu in the normal and primary invasive breast cancer tissues [26].

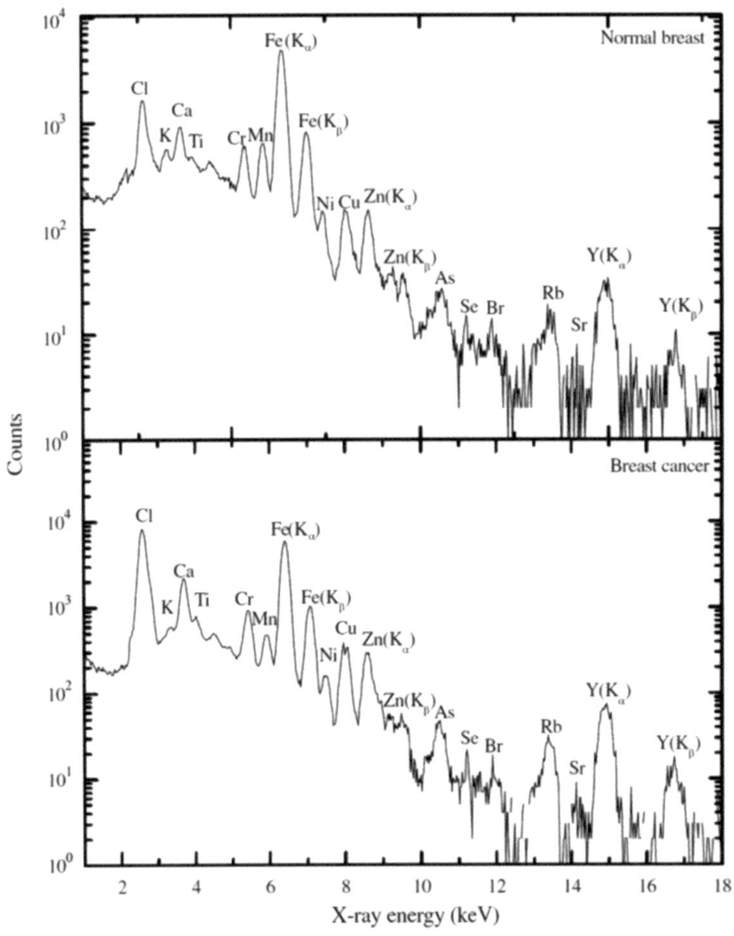

Figure 4.
Normal and abnormal breast cancerous tissue obtained from PIXE spectrum [37].

Breast tissue samples are generally collected from mastectomies, lumpectomies and breast reduction surgery [27–29]. Since, among the healthy population, a significant variation of trace elemental levels is found, generally healthy tissues from areas distant to malignant tissues/neoplastic are also collected from the same individual for comparison. For measurements like SIXRF to be performed directly on the wet tissues, collected tissue Samples are washed with milli-Q water to remove any stain of blood and are stored in formalin (10% formaldehyde in water) at room temperature or kept frozen at <−40°C until the analysis [27, 29, 30]. For EDXRF measurements, tissue samples are lyophilised for at −60°C and low pressure, approximately 10^{-1} atm, where the low temperature ensures the retention of volatile elements like Arsenic (As) and Mercury (Hg). These dried tissue samples are milled by freezer mill cooled by liquid nitrogen and pressed into pellets without any further chemical treatment or additive [7]. For techniques TXRF, requiring thin sections, tissue samples are cut into small pieces and excised into cylindrical pieces after cooling with liquid nitrogen. The excised tissue is then cut with a cooled microtome to get thin sections. Thin section is positioned in the centre of the sample carrier which is made non-hydrophobic by means of alcoholic silicon solution and then dried. Thin sections are spiked with internal standard solution and are again dried [31]. Various researchers have also used breast cancerous blood and blood serum for trace elements determination. For determination of trace element in human blood samples 4 ml of blood was mixed with 1350 mg cellulose and freeze dried. A fine powder of this mixture was pressed into pellets and used for WD-XRF measurements [32]. Blood serum samples are also lyophilized, and a small pellet is made which has been used for SIXRF in study [33]. For PIXE analysis of these serum samples, to make the sample conducting, graphite powder is also added and and pelletized [34]. **Table 1** shows the trace elements levels reported in the literature for normal and abnormal human breast cancer using XRF techniques. The values are given in μg/g.

2.1 Quantification of trace elements

2.1.1 WDXRF

The quantitative elemental analysis of different blood samples (normal and abnormal) was performed by using a commercial WDXRF spectrometer equipped with different anodes X-ray tubes (Rh/Ag/W), a gas flow proportional (FP) counter and a scintillation counter (SC) as photon detectors. The mass concentrations of different elements present in samples were determined using advanced software package available with the spectrometer. The intensity of X-ray lines for the specific element is determined with Lachance-Traill method which is defined as

$$I_g = D_g C_g \times \left(1 + \sum_{w \neq g} R_{gw} . m_w \right) \quad (1)$$

where, I_g corresponds to the intensity of the specific element g, C_g refers to the measured concentration of the corresponding element, D_g is the instrumental calibration coefficient for the given element. The term m_w denotes the concentration of the other element w and R_{gw} is the inter-element matrix coefficients. This software enables to evaluate the accurate concentrations of different elements ranging from Be to U present in unknown samples by incorporating corrections due to the matrix effects. The data acquisition time for each target was kept usually as ~20 minutes to collect good statistics under different X-ray peaks arising from different elements present in each sample.

Elements	Na	Mg	P	S	Cl	K	Ca	Cr	Fe	Cu	Zn	As	Se	Sr	Ref.	Analytical Technique
Normal									11.6	0.3	2.7					
Abnormal									18.8	0.9	6.5				[27]	SR-XRF
Normal						438			14.1	0.3	2.9					
Abnormal						1112			21.7	1	6.9				[29]	SR-XRF, EDXRD
Normal						210			7	0.29	1.8					
Abnormal						1032			18.8	0.95	7.7				[29]	SR-XRF, EDXRD
Normal							726		9.0	0.6	3.8					
Abnormal							2400		25.3	1.8	12.9				[30]	SR-XRF
Normal					3999.9	1381.0	157.1	32.0	299.5	42.0	56.2	2.57	0.7	6.7		
Abnormal					6815.4	1550.8	480.2	52.7	376.3	60.7	126.2	4.12	1.3	13.7	[37]	PIXE
Normal				295		17	153		5	2	6					
Abnormal				3240		165	1270		98	14	39	—	—	—	[7]	TXRF
Normal						112	970		32	35	31			6		
Abnormal						360	1530		147	27	64			8	[7]	EDXRF
Normal						1975	1259	3.2	28.1	14.8	19.2	0.06	1.7			
Abnormal						2175	1320	2.9	56.3	19.5	9.8	0.07	0.8		[33]	SR-XRF
Normal								18.6	291	24.6	38.5	0.9	2.5			
Abnormal								10.4	355	32.3	13	0.7	1.5		[34]	PIXE
Normal	1437.2	66.7	354.5	1618.6	4805.5	3164	131.5		1231.9	19.9	30.4					
Abnormal	2494.1	83	590.7	2631.6	7372.9	4309.5	294.8		1885	36.1	57.5				[32]	WDXRF
Normal				1908		3.0	100.6		3.4	1.7	16.9					
Abnormal				2132		4.4	127.4		8.7	3.3	11.4				[39]	PIXE

Table 1.
Trace elements levels reported in the literature for normal and abnormal human breast cancer using XRF techniques. the values are given in µg/g.

2.1.2 For TXRF

The qualitative and quantitative analysis of this technique requires an internal standard (U) solution of known concentration with associating its concentration relative to the net intensities found in the analyte by using the following equation

$$C_G = \frac{\frac{N_P}{S_P}}{\frac{N_R}{S_R}} C_U \tag{2}$$

Where, C_G/ C_U are the concentration of the analyte / known internal standard solution. The term N_P/ C_U refer to the net intensity of the given analyte and S_R/C_U correspond to the relative sensitivity of the analyte/known standard solution.

2.1.3 For PIXE

The multi-element analysis of unknown samples can be obtained by using available commercial peak fitting software GUPIXWIN (University of Guelph, Ontario, Canada) which is given as

$$P_z = P_t(Z) \frac{4\pi mzZ_x e}{\sigma_x^z d\varphi eff_p QN_A} \tag{3}$$

The term P_t *(Z)* is the counts of the X-ray fluorescent spectra for any atomic number (Z), $d\,\varphi$, eff_p, Q, Z_x and N_A are the solid angle of the X-ray detector, efficiency, integrated projectile charge, projectile charge, and Avogadro's number, respectively.

2.1.4 For EDXRF

In this type of technique, the elemental concentration of unknown samples is evaluated by

$$m_i = \frac{N_{jk}}{I_o G \in_{jk} \beta_{jk} \sigma_{jk}} \tag{4}$$

Where, m_k denotes the concentration of kth element present in the sample, N_{jk} is the net rate for the jth group of X-rays of kth element, $\in_{jk}$ is the detector efficiency for the jth X-ray energy of the kth element, I_oG is the intensity of the exciting radiation incident on the sample visible to the detector, σ_{jk} is the X-ray fluorescence cross section of the jth X-rays of the kth element at the incident photon energy, and β_{jk} is the self-absorption correction factor which accounts for the absorption of the incident and the emitted characteristic X-rays lying under the ith peak of kth element within the target.

3. Role of trace elements

In the past, various researchers have reported a high level of iron in human breast cancer tissues as compared to the normal one [27–30, 35–38]. Elevated iron levels are also reported in in blood [32, 34] and scalp hair [39] of breast cancer patients using PIXE and WDXRF. Iron (Fe) plays a vital role in the human body and is an essential element. Normally, the human body contains 4–5 g of iron out of which 1 g is stored in the liver and spleen. The main component forms of iron, hemoglobin and myoglobin help in the growth of cells. The low and high dosage of iron cause various diseases like heart diseases, diabetes, anemia, cancer, listlessness, stomatitis, etc. and promote cancer which damage the tissues and convert hydrogen peroxide to free radical ions via Fenton and Haber-Weiss type reactions. These free radical ions cause DNA strand breaks, sister-chromatid and initiate lipid peroxidation which promotes the growth of cancer [40, 41].

An Element like copper (Cu) is involved in multiple biological processes which promote tumor growth. As far as the role of Cu in breast cancer is concerned, the picture comes out to be rather wavy. Morton K. Schwartz in his research also reported the role of trace elements in cancer [42]. The author mentioned that the Cu level in breast cancer tissue was greater than the normal one. Studies [27, 29, 30, 37, 38] using various techniques are in well agreement with past results. Many studies showed that the level of Cu and in blood serum [32, 34] and in hairs [39] of breast cancer patients are higher as compared to the normal one. The daily requirement of Cu intake is about 2 mg/day and heavy dose ingestion causes various diseases. The role of Cu and its concentration is well explained by many workers [43–45]. The toxicity and abnormal level of Cu element and metabolism processes present in the human blood cause the formation of blood vessels which further results in various types of cancer like breast, brain, gladder, etc. The extra formation of blood vessels in the human body is called Angiogenesis. It plays a vital role in the evolution of cancer cells inside the body. Since blood flows in the whole body, these cells also require blood for their growth, so it gives chemical signals to stimulate Angiogenesis. The matrix metalloproteinase (MMP) family of enzymes degrades the basement membrane and extracellular matrix of tissue inhibitors of metalloproteinase (TIMP). Under critical condition both MMP and TIMP imbalance the tissue and activate angiogenesis which caused breast cancer [46].

For zinc (Zn) element, the concentration of Zn in breast cancer case is slightly large as compared to the normal one. Similar trends were also reported by many researchers by using different techniques and methods [29, 30, 37, 38, 47, 48]. Depressed Zn levels are found in blood sera of breast cancer patients using PIXE [34] but higher levels in blood are found in [32] using WDXRF. This might be understood in terms of biochemical and histological differences between cancerous and normal blood. Like other elements, Zn also plays an important role in the biological, physiological and metabolic processes of the human blood. It is also obligatory for the formation and common function of the cell membrane. Toward the role in cancerous blood or tissues, the statement about the Zn is contradictory. Earlier reports suggested that the abnormal level of Zn leads to carcinogenesis [49]. These inconsistent annotations suggest that the role of Zn may vary from one to another organ depending on various factors like age group, lifestyle, environment changing and diet etc. However, in breast cancer case, level of Zn increases in cancer case rather than the normal case which is explained earlier. Lee et al. [50] also gave evidence on the behavior of Zn in human normal and cancer tissues. In their research, they suggested that the altered Zn homeostatic in breast cancer tissue is responsible for the increased level of Zn in the cancerous case which possibly leads to the growth of breast tumor. Zn is an important trace metal being a cofactor for more than 300 enzymes, and contributes to cellular signaling, proliferation, homeostasis, apoptosis [51, 52]. It is also a structural component of more than ~3000 proteins including metallothionein's, zinc transporters, p53 tumor suppressor and matrix metalloproteases which are involved in carcinogenesis and cancer progression [53]. In particular, p53 activation is important for apoptosis and cell cycle arrest in breast cancer case and protects women from it. Transcription factors, e.g. nuclear factor-kappa (NF-κB) is activated in the breast cancer and leads to a more aggressive phenotype. Association of Zinc to breast cancer cell inhibits NF-κB [54]. Generally, the level of Zn in breast cancer cases is more. The reason behind them is that Zn is necessary for the cell production in a region adjacent to the tumor due to the presence of MMPs or tissue inhibitors of metalloproteinase. Similarly, the observed correlation between Cu and Zn shows that the ratio of Cu/Zn is more in cancerous blood as compared to the normal blood [34, 48].

The element chromium (Cr) is also responsible for the formation of breast cancer in human body. Earlier reports suggested that level of trace elements of Cr in breast cancer tissues were significantly higher than the normal ones [37]. But in the research article given by Sarita et al. [34] using PIXE technique, the level of Cr in blood sera was found to be lower in breast cancer case as compared to the normal one. As we know that carcinogenic property of the trace elements depends mainly upon the factors like oxidation states and their chemical structure. In case of Cr element, the hexavalent chromium compounds Cr (VI) are more toxic than trivalent form Cr (III). Cr (VI) is easily absorbed by the body cells and then reduced to the trivalent form i.e. Cr (III). This reduction generates free active oxygen radical as well as glutathionyl radicals which further produces genotoxic effects which are responsible for the formation of breast cancer [55].

Chlorine (Cl) is an invasive chemical used in daily life products. It plays an important role in balancing the body cells and helps to digest the food. In the earlier study, the trends of the Cl element in human cancer tissue and blood were higher than the normal ones [32, 37]. Generally, Cl is present in extracellular fluid. Combination of Cl with organic compounds present in water or soil forms organo-chlorine. These are widely used as standard pesticides e.g. DDT, DDD, Isobenzan, Dicofol, Dieldrin, Eldrin, Lindane, BHC, etc. Carcinogenic and the weak estrogenic and anti-estrogenic hormonal effects of many organochlorines and their hydroxylated metabolites have led many researchers to hypothesize that they increase the risk of breast cancer in humans [56]. These are responsible for impairment or suppression of cell mediated immunity [57], mimicking androgenous hormones and modulating their as well as estrogen hormones metabolism prompting breast cancer development [58].

Since Ca is the major constituent of breast tissue calcification in the form of calcium hydroxyapatite, it has of vital importance in breast cancer. For the blood clotting mechanism, Ca plays an important role as factor IV. The Ca^{2+} cation is involved in various electrochemical mechanisms in the body like neutralization of charge, emulsion stabilization, free energy supply to the body cell. Magalhães et al. [38] mentioned the increased behavior of Ca in their research with TXRF technique. The same trend was also given by various workers with different methods and techniques [30, 37, 59, 60]. Higher Ca levels are found in the scalp hair and blood of breast cancer patients using PIXE and WDXRF in [32, 39]. Cationic calcium homeostasis in the plasma is tightly maintained through absorption, excretion and secretion and storage in bone. High level of calcium in the blood serum, known as hypercalcaemia, is associated with different types of cancer like breast cancer, lung cancer and myeloma, etc. These types of cancer result Ca to leak out from bones to the blood making it heavier than the normal one. This generally happens to the people suffering from cancer in the advance stage. Various analyses like meta-analysis and dose response analysis have been done to establish fact that there is a significant relation between Ca intake and breast cancer risk [61]. In these studies, the researchers suggested the dietary, lifestyle and intake dose of Ca which affect the human body. Having a complex nature, Ca intake is known to be inversely associated to breast cancer risk significantly in pre and postmenopausal women [62]. With rising cases of breast cancer being reported in the literature, the trace elements like phosphorus (P), sulfur (S), potassium (K) and sodium (Na) and their role should be of greater concern. From the literature, it has been seen that proportion of these elements in breast cancer human blood is slightly higher with respect to the normal human blood. Earlier reports found that the abnormal level of phosphorus may influence breast cancer [32]. Some studies reported a statistically elevated P content in breast tissue compared to normal one using TXRF [31, 38, 48]. The number of in-vitro studies suggested that for cell growth in human body, the inorganic phosphate (Pi) and phosphate are two responsible terms. They both act like a mitogen. This elevated value of Pi promotes cell prolific microenvironment which causes breast tumor. Another study by Wulaningsih and his co-workers showed that when the content of phosphorus along with calcium enters in the human body, it increases the estrogens level which promotes the growth of breast cancer [63, 64].

In a view of sulfur element role, it has been seen from the past studies that the concentration of the sulfur increased in cancerous blood as well as in cancerous tissue than normal ones [31, 32, 38]. Sulfur plays an important role in cell renewal and enables transferring of oxygen from cell membrane. It is widely used in biological processes which act as both fuel and respiratory materials for the human body. Sulfur has an important role in chemotherapy to reduce the size of the breast tumor which uses sulfur containing drugs like Docetaxel, Paclitaxel, Taxanes, Eribulin, etc. [65]. From the past studies, it has been seen that the sulfur is commonly used in some form like organic sulfur to treat cancer which is useful in anti-cancer therapy. The research also claims that the sulfur containing compounds work like an anti-cancer reagent which kills the cancer cells without affecting normal and healthy cells present in the whole body system. Furthermore, organic sulfur compounds like amino acid, methyl-sulfonyl methane and diallyl sulphide, etc. have powerful anti-cancer effect against breast cancer [66].

Elements like potassium (K) and sodium (Na), they both also play a key role in the biological processes present in the human body system. The literature clearly shows that the value of both potassium and sodium in human breast cancer cell and blood increases with respect to normal one [6, 29, 32, 38]. Earlier views on potassium element clearly show that the concentration in human affected from breast cancer is not significantly different from the normal one [37]. This twin behavior of potassium element might depend upon many factors like eating/drinking lifestyle, environment behavior, sample preparation etc. Also, it has been concluded that most of the research has been done on the cancer tissues which are generally not homogeneous. So, for multi-elements detection system and for obtaining better results, samples must be homogeneous. In order to understand the role of potassium element in the cancerous blood, we know that it acts as an electrolyte and present mainly in the form of k^+ ion (cation) inside the human blood. Acid–base and water balance in the tissues and blood is maintained with the help of K^+ in the human body. The role of K^+ in regulating tumor cell proliferation and as anti-apoptotic and pro-apoptotic agent is well established [67]. On its combination with ascorbic acid, the inhibitory effect on the survival of breast cancer cell lines has been observed. Further details are given in Ref. [68]. On the other hand, in the case of sodium element, we clearly have seen from the past studies that the concentration of sodium element increases in case of breast cancer blood of human body [32]. The role of sodium (compound form) in cancerous blood is also a big concern. Since sodium is also present in the human blood in the cation form, it also plays an important role in the metabolism processes in the human body. Researchers reported the activity of Na^+/K^+ adenosine tri-phosphate (ATP) which clearly explain the difference between the concentration of Na + and k + cation in cancerous cells and normal one [69]. Higher Na^+/K^+ ratios have been reported in cancerous blood using WDXRF [32].

The recent research says that the sodium compound form i.e. sodium bicarbonate can be treated as an anti-cancer agent. The research proved that the use of sodium bicarbonate has been beneficial for breast and prostate cancer etc. Most of the work was done on the mice.

Magnesium (Mg) is the second most abundant cation in the body which acts as an activator in about 300 enzymatic reactions. These reactions include conversion of ATP to ADP for cell energy metabolism, DNA replication and repair, protein synthesis, inflammation, proliferation, cell cycle control and apoptosis [70, 71]. Most of these factors are related to carcinogenesis. Progression of breast cancer is also related to Mg where in the proliferative phase of the disease, the neoplastic cells causes an influx of it and hence there is an increase in the intracellular concentration of Mg disturbing its homeostasis [72]. Several studies have investigated the direct association of high magnesium intake and breast cancer risk and an inverse relation has been found. Indirectly, higher Mg intake lowers the C Reactive Protein (CRP) level and decreases the breast cancer risk [73]. Some earlier reports suggested that a significantly low Mg level in serum found with respect to normal tissue [74] while high Mg levels are found in cancerous blood using WDXRF in [32]. It is also observed that the concentrations of magnesium in all blood samples are opposite to the concentration of calcium. Both calcium and magnesium compete for the transporter transient receptor potential melastatin 7 (TRPM7) for their absorption in the tumor membrane. A negative feedback mechanism regulates the level of both. It has been seen that reduced level of magnesium decreases the Mg-ATP levels inside the cell which leads to increase the Ca-ATP levels of the cell and this intracellular Ca increase leads to increased cell proliferation. Ca/Mg ratio is elevated in breast cancer case since Ca concentrations are increased while concentration of Mg is decreased [75].

Selenium (Se) is the important trace mineral for the human body. It is believed that all the enzymes present in the human body system are selenium dependent. The abnormal level of selenium causes many diseases. As far as a disease like breast cancer is concerned, the picture of selenium is not clear. Different studies give different views on the role and presence of selenium in the human body. Using TXRF technique, Magalhães et al. [38] mentioned that the concentration of selenium element increased in breast cancer tissue as compared to the normal one. The observed high level of selenium in breast cancer tissue was also reported in the earlier studies also [37]. Low Se levels are found in the blood sera samples in [34]. In accordance with the hypothesis, finding suggested that the selenium worked as an immune enhancing and antioxidant to reduce the breast cancer [76]. Lifestyle, eating/drinking habits, environment, etc. are deciding factors because most of the selenium found in the human body is from eating/drinking habits etc. An inverse relationship exists between Selenium intake and risk of breast cancer [77]. Also, it might be guessed that it acts like an anti-cancerous agent. We know that selenium is generally absorbed in the body in the form of L-selenomethionine. Earlier studies confirmed that in MCF-7 breast cancer, the therapeutic effect of methyl selenocysteine combined with tamoxifen and imidoselenocarbamate considered as an antitumour agent reduces the growth of breast cancer [78, 79]. Combs et al. [80] in their research article suggested that selenium works as anti-oxidant effects via Glutathione peroxidise (GSH-Px) which further protects the body from damaging effect of free radicals.

Arsenic (As) is one of the most toxic elements found in nature. The main source of arsenic coming in our body is from eating/drinking habits, soil and from plants etc. However, its mechanism and role are not well explained but the epidemiological evidence in the past literature showed the toxicity of arsenic from drinking water causes different types of cancer. From the literature, it has been seen that the level of arsenic is higher in breast cancer tissue as compared to others by using XRF techniques [37]. Lower serum As concentrations are found in [34]. The reason behind this is that the arsenite generates the effect of estradiol and induces ROS growth, DNA damage and increases c-Myc and heme oxygenase (HO-1) protein levels which lead to tumor cell proliferation and increase in the estrogens level in the body and causes breast cancer in MCF-7 cells. The c-Myc is one of the most commonly activated genes present in advance stages of breast cancer. The study shows that arsenite present in breast cancer MCF-7 cells increases the c-Myc and HO-1 level which results in DNA damage. The high increases in c-Myc and HO-1 level further deactivate the p53 gene and affect the metabolism and biological process if the body results in breast cancer [81]. On the other hand, arsenic trioxide has an antiproliferative effect on human breast cancer MCF-7 cells due to reduction of HERG channels and activation of caspase-3. Generally, HERG belongs to multi-genetic family of voltage gate k + channels and present mostly in the tumor cells, not in normal cells of the human body system [82].

For Strontium, the Department of Health and Human Services determined that stable isotopes of strontium do not play any role in cancer. Its radioactive isotopes ^{89}Sr and ^{90}Sr are important for breast cancer. Earlier studies mentioned the higher level of strontium in breast cancer tissue as compared to the normal tissue [37]. However, the IARC clearly suggested the carcinogenic behavior of radioactive strontium (^{90}Sr) which may cause cancer. The reason is that when ^{90}Sr enters in the body it gets mostly attached on the surface of the bone and soft tissue itself. Due to high dose and radioactive decay property, it combines with the blood or tissue and damages the DNA structure. In the case of ^{89}Sr, the previous study showed that it is more beneficial in breast cancer patients with metastatic bone pain and have similar metabolism function as that of calcium [83].

Acknowledgements

Author (H. S. Kainth) acknowledges the financial support received from University Grant Commission (UGC) New-Delhi, grant number F.4-2/2006 (BSR)/PH/18-19/0084, in the form of D. S. Kothari post-doctorate fellowship.

Author details

Harpreet Singh Kainth[1*], Deeksha Khandelwal[2], Ranjit Singh[3], Gurjeet Singh[4] and Sanjiv Puri[1]

1 Department of Basic and Applied Sciences, Punjabi University, Patiala, India

2 Inter University Accelerator Centre, Aruna Asaf Ali Marg, New Delhi, India

3 Department of Radiotherapy, PGIMER, Chandigarh, India

4 Department of Physics, Punjabi University, Patiala, India

*Address all correspondence to: harpreet.2january@gmail.com

References

[1] Bray F, Ferlay J, Soerjomataram I, Siegel RL, Torre LA, Jemal A. Global cancer statistics 2018: GLOBOCAN estimates of incidence and mortality worldwide for 36 cancers in 185 countries. CA: a cancer journal for clinicians. 2018;68(6):394-424. DOI: 10.3322/caac.21492

[2] Brinton LA, Gaudet MM, Gierach GL. Breast cancer. In: Thun MJ, Linet MS, Cerhan JR, Haiman CA, Schottenfeld D, editors. Cancer Epidemiology and Prevention. 4th ed. New York: Oxford University Press;2018. p. 861-888. DOI: 10.1093/oso/9780190238667.001.0001

[3] Liou G-Y, Storz P. Reactive oxygen species in cancer. Free Radical Research. 2010; 44(5):1-31. DOI: 10.3109/10715761003667554

[4] Reddy S, Charles M, Naga Raju G, Byreddy S, Reddy T, Lakshmi P, Vijayan, V. Trace Elemental Analysis of Cancer-Afflicted Intestine by PIXE Technique. Biological trace element research. 2004; 102:265-282. DOI: 10.1385/bter:102:1-3:265

[5] Czarnowski D V, Denkhaus E, Lemke K. Determination of trace element distribution in cancerous and normal human tissues by total X-ray fluorescence analysis. Spec. Acta B. 1997; 52(7):1047-1052. DOI: 10.1016/S0584-8547(96)01625-4

[6] Milde D, Altmannová K, Vysloužil K, Stužka V. Trace Element Levels in Blood Serum and Colon Tissue in Colorectal Cancer. Chemical Papers. 2005; 59(3): 157-160. Corpus ID: 17410947

[7] Magalhães T, Bohlen A V, Carvalho M L, Becker M. Trace elements in human cancerous and healthy tissues from the same individual: A comparative study by TXRF and EDXRF. Spec. Acta Part B. 2006; 61:1185-1193. DOI: 10.1016/j.sab.2006.06.002

[8] Benninghoff L, Czarnowski D V, Denkhaus E, Lemke K. Analysis of human tissues by total reflection X-ray fluorescence. Application of chemometrices for diagnostic cancer recognition. Spec. Acta Part B. 1997; 52:1039-1046. DOI: 10.1016/S0584-8547(96)01626-6

[9] Hernandez-Caraballo E A, Marco-Parra L M. Direct analysis of blood serum by total reflection X-ray fluorescence spectrometry and application of an artificial neutral network approach for cancer diagnosis. Spec. Acta Part B. 2003; 58:2195-2201. DOI: 10.1016/j.sab.2003.07.003

[10] Dyson N. X-rays in Atomic and Nuclear Physics. 2nd ed. Cambridge: Cambridge University Press; 1990. DOI: 10.1017/CBO9780511470806

[11] Jenkins R, Gould R W, Gedcke D. Quantitative X-ray spectrometry. CRC Press; 1995. DOI: 10.1201/9781482273380

[12] Stand-zenieks P, Selin E. Background reduction of X-ray fluorescence spectra in a secondary target energy dispersive spectrometer. Nucl. Inst. Methods Phys. Res., B Beam Interact. Mater. Atoms. 1979; 165: 63-70. DOI: 10.1016/0029-554X(79)90308-2

[13] Custódio P, Carvalho M L, Nunes F, Trace elements determination by energy dispersive X-ray fluorescence (EDXRF) in human placenta and membrane: A comparative study. Anal. Bional. Chem. 2003; 375:1101-1106. DOI: 10.1007/s00216-003-1765-9

[14] Cesareo R, Castellano A, Buccolieri G, Quarta S, Marabelli M, Santopadre P, Leole M, Brunetti A, Portable equipment for energy dispersive X-ray fluorescence analysis of Giotto's frescoes in the Chapel of the Scrovegni, Nucl. Instrum. Methods

Phys. Res., B Beam Interact. Matter. Atoms. 2004; 213:703-706. DOI: 10.1016/S0168-583X(03)01758-0

[15] Wobrauschek P, Total reflection X-ray fluorescence analysis-a review, X-Ray Spectrom. *2007;36:*289-300. DOI: 10.1002/xrs.985

[16] Tsuji K, Emoto T, Matsuoka Y, Nagamura T, Ding X, Micro-xrf instrument developed in combination with atomic force microscope. Adv. X Ray Anal. 2005; 48:221-228. DOI: 10.1154/1.1913724

[17] Buzanich G, Wobrauschek P, Streli C, Markowicz A, Wegrzynek D, Chinea-Cano E, Bamford S. A portable micro-X-ray fluorescence spectrometer with polycapillary optics and vacuum chamber for archaeometric and other applications. Spectrochimica Acta Part B: Atomic Spectroscopy. 2007;62(11):1252-1256. DOI: 10.1016/j.sab.2007.08.003

[18] Lopes R T, Lima I, Pereria G R, Perez C A. Synchrotron radiation X-ray micro fluorescence techniques and biological applications. Pramana-J. Phys. 2011; 76:271-279. DOI: 10.1007/s12043-011-0043-1

[19] Johansson SA, Campbell J L. PIXE: A novel technique for elemental analysis. John Wiley & Sons, Chichester, UK, 1988

[20] Ryan C G. Quantitative Trace Element Imaging Using PIXE and the Nuclear Microprobe. John Wiley & Sons, North Ryde, Australia, 2001

[21] Ryan C G, Cousens D R, Sie S H, Griffin W L, Suter G F, Clayton E. Quantitative pixe microanalysis of geological matemal using the CSIRO proton microprobe. Nucl. Instru. Meth. Phy. Res. B. 1990; 47:55-71. DOI: 10.1016/0168-583X(90)90047-X

[22] Yano J, Yachandra VK. X-ray absorption spectroscopy. Photosynthesis research. 2009;102(2-3):241. DOI: 10.1007/s11120-009-9473-8

[23] Błachucki W, Czapla-Masztafiak J, Sá J, Szlachetko J. A laboratory-based double X-ray spectrometer for simultaneous X-ray emission and X-ray absorption studies. Journal of Analytical Atomic Spectrometry. 2019;34(7):1409-1415. DOI: 10.1039/C9JA00159J

[24] Koningsberger DC, Prins R. X-ray absorption: principles, applications, techniques of EXAFS, SEXAFS, and XANES. United States: John Wiley and Sons. 1988

[25] Penner-Hahn JE. X-ray absorption spectroscopy. e LS. 2001. DOI: 10.1038/npg.els.0002984

[26] Al-Ebraheem A, Goettlicher J, Geraki K, Ralph S, Farquharson MJ. The determination of zinc, copper and iron oxidation state in invasive ductal carcinoma of breast tissue and normal surrounding tissue using XANES. X-Ray Spectrometry. 2010;39(5):332-337. DOI: 10.1002/xrs.1272

[27] Geraki K, Farquharson MJ, Bradley DA. Concentrations of Fe, Cu and Zn in breast tissue: a synchrotron XRF study. Physics in Medicine & Biology. 2002;47(13):2327-2339. DOI: 10.1088/0031-9155/47/13/310

[28] Geraki K, Farquharson MJ, Bradley DA. X-ray fluorescence and energy dispersive x-ray diffraction for the quantification of elemental concentrations in breast tissue. Physics in Medicine & Biology. 2003;49(1):99-110. DOI: 10.1088/0031-9155/49/1/007

[29] Geraki K, Farquharson MJ, Bradley DA, Hugtenburg RP. A synchrotron XRF study on trace elements and potassium in breast tissue. Nuclear Instruments and Methods in Physics Research Section B: Beam Interactions with Materials and Atoms. 2004;213:564-568. DOI: 10.1016/S0168-583X(03)01672-0

[30] Silva MP, Tomal A, Perez CA, Ribeiro-Silva A, Poletti ME. Determination of Ca, Fe, Cu and Zn and their correlations in breast cancer and normal adjacent tissues. X-Ray Spectrometry: An International Journal. 2009;38(2):103-111. DOI: 10.1002/xrs.1126

[31] Magalhaes T, Becker M, Carvalho ML, Von Bohlen A. Study of Br, Zn, Cu and Fe concentrations in healthy and cancer breast tissues by TXRF. Spectrochimica Acta Part B: Atomic Spectroscopy. 2008;63(12):1473-1479. DOI: 10.1016/j.sab.2008.10.014

[32] Singh R, Kainth HS, Prasher P, Singh T. Trace elemental analysis of human breast cancerous blood by advanced PC-WDXRF technique. Nuclear Instruments and Methods in Physics Research Section B: Beam Interactions with Materials and Atoms. 2018;419:44-48. DOI: 10.1016/j.nimb.2018.01.029

[33] Naidu BG, Sarita P, Raju GN, Tiwari MK. Multivariate analysis of trace elemental data obtained from blood serum of breast cancer patients using SRXRF. Results in Physics. 2019;12:673-680. DOI: 10.1016/j.rinp.2018.12.020

[34] Sarita P, Naga Raju G, Pradeep A, Rautray T, Seetharami Reddy B, Bhuloka Reddy S, Vijayan V. Analysis of trace elements in blood sera of breast cancer patients by particle induced X-ray emission. Journal of Radioanalytical and Nuclear Chemistry. 2012;294(3):355-361. DOI: 10.1007/s10967-011-1505-0

[35] Kwang-Hoong N G, Bradley D A, Lai-Meng L, Elevated trace element concentrations in malignant breast tissues. The Br. J. Radiol. 1997;70:375-382. 10.1259/bjr.70.832.9166074

[36] Kuo H W, Chen S F, Wu C C, Chen D R, Lee J H. Serum and tissue trace elements in patients with breast cancer in Taiwan. Biol. Trace Elem. Res. 2002; 89 (1): 1-11. DOI: 10.1385/BTER:89:1:1

[37] Naga Raju G, Sarita P, Kumar M R, G.A.V.R. Murty, Reddy B S, Lakshminarayana S, Vijayan V, Lakshmi P V B R, Gavarasan S, S.B. Reddy. Trace elemental correlation study in malignant and normal breast tissue by PIXE technique. Nucl. Inst. and Meth. In Phys. Res. B 2006;247 (2): 361-367.

[38] Magalhães T, Carvalho ML, Von Bohlen A, Becker M. Study on trace elements behaviour in cancerous and healthy tissues of colon, breast and stomach: Total reflection X-ray fluorescence applications. Spectrochimica Acta Part B: Atomic Spectroscopy. 2010;65(6):493-498. DOI: 10.1016/j.sab.2010.04.001

[39] Kabiri Z, Kakuee O, Fathollahi V, Stout B. Trace element abnormalities in the scalp hair of breast cancer patients. International Journal of PIXE. 2014;24(01n02):49-58. DOI: 10.1142/S0129083514500065

[40] Weinberg ED. The role of iron in cancer. European journal of cancer prevention: the official journal of the European Cancer Prevention Organisation (ECP). 1996;5(1):19-36. PMID: 8664805

[41] Toyokuni S. Role of iron in carcinogenesis: cancer as a ferrotoxic disease. Cancer science. 2009;100(1):9-16. DOI: 10.1111/j.1349-7006.2008.01001.x

[42] Schwartz M K. Role of trace elements in cancer. Cancer Res. 35 (11 Pt.2) (1975) 3481-3487. PMID: 1104155

[43] Sky-Peck H H. Trace metals and Neoplasia. Clin. Physiol. Biochem. 1986; 4:99-111. PMID: 3514058

[44] Hrgovcic M, Tessmer CF, Thomas FB, Ong PS, Gamble JF, Shullenberger CC. Serum copper

observations in patients with malignant lymphoma. Cancer. 1973;32(6):1512-1524. DOI: 10.1002/1097-0142(197312)32:6<1512::AID-CNCR2820320631>3.0.CO;2-P

[45] (Ed.: H. Sigel), Metal ions in Biological Systems: Carcinogenicity and Metal Ions, Marcel Dekker, Inc.: New York, 1980.

[46] Quintero-Fabián S, Arreola R, Becerril-Villanueva E, Torres-Romero JC, Arana-Argáez VE, Lara-Riegos J, Ramírez-Camacho MA, Alvarez Sanchez ME. Role of matrix metalloproteinases in angiogenesis and cancer. Frontiers in oncology. 2019;9:1370. DOI: 10.3389/fonc.2019.01370

[47] Mulware SJ. Comparative trace elemental analysis in cancerous and noncancerous human tissues using PIXE. Journal of biophysics. 2013;2013. 10.1155/2013/192026

[48] Carvalho ML, Magalhães T, Becker M, Von Bohlen A. Trace elements in human cancerous and healthy tissues: A comparative study by EDXRF, TXRF, synchrotron radiation and PIXE. Spectrochimica Acta Part B: Atomic Spectroscopy. 2007;62(9):1004-1011. DOI: 10.1016/j.sab.2007.03.030

[49] Kew M C, Mallett R C. Hepatic zinc concentrations in primary cancer of the liver. Brit. J. Cancer. 1974; 29 (1): 80-83. 10.1038/bjc.1974.11

[50] Lee R, Woo W, Wu B, Kummer A, Duminy H, Xu Z. Zinc accumulation in N-methyl-N-nitrosourea-induced rat mammary tumors is accompanied by an altered expression of ZnT-1 and metallothionein. Experimental Biology and Medicine. 2003;228(6):689-696. PMID: 12773700

[51] Franklin RB, Costello LC. The important role of the apoptotic effects of zinc in the development of cancers. Journal of cellular biochemistry. 2009;106(5):750-757. DOI: 10.1002/jcb.22049

[52] Maret W. Zinc in cellular regulation: The nature and significance of "zinc signals". International Journal of Molecular Sciences. 2017;18(11):2285. DOI: 10.3390/ijms18112285

[53] Takatani-Nakase T. Zinc transporters and the progression of breast cancers. Biological and Pharmaceutical Bulletin. 2018;41(10):1517-1522. DOI: 10.1248/bpb.b18-00086

[54] Grattan BJ, Freake HC. Zinc and cancer: implications for LIV-1 in breast cancer. Nutrients. 2012;4(7):648-675. DOI: 10.3390/nu4070648

[55] Sugiyama M. Role of physiological antioxidants in chromium (VI)-induced cellular injury. Free Radical Biology and Medicine. 1992;12(5):397-407. 10.1016/0891-5849(92)90089-y

[56] Raaschou-Nielsen O, Pavuk M, LeBlanc A, Dumas P, Weber JP, Olsen A, Tjønnland A, Overvad K, Olsen JH. Adipose organochlorine concentrations and risk of breast cancer among postmenopausal Danish women. Cancer Epidemiology and Prevention Biomarkers. 2005;14(1):67-74. PMID: 15668478

[57] Vineis P, D'Amore F, Working Group on the Epidemiology of Hematolymphopoietic Malignancies in Italy. The role of occupational exposure and immunodeficiency in B-cell malignancies. Epidemiology. 1992:266-270. http://www.jstor.org/stable/3703163

[58] Calaf GM, Ponce-Cusi R, Aguayo F, Muñoz JP, Bleak TC. Endocrine disruptors from the environment affecting breast cancer. Oncology Letters. 2020 Jul 1;20(1):19-32. DOI: 10.3892/ol.2020.11566

[59] Rizk SL, Sky-Peck HH. Comparison between concentrations of trace elements in normal and neoplastic human breast tissue. Cancer research. 1984;44(11):5390-5394. PMID: 6488192

[60] Cui Y, Rohan TE. Vitamin D, calcium, and breast cancer risk: a review. Cancer Epidemiology and Prevention Biomarkers. 2006;15(8):1427-1437. DOI: 10.1158/1055-9965.EPI-06-0075

[61] Hidayat K, Chen GC, Zhang R, Du X, Zou SY, Shi BM, Qin LQ. Calcium intake and breast cancer risk: meta-analysis of prospective cohort studies. British Journal of Nutrition. 2016;116(1):158-166. DOI: 10.1017/S0007114516001768

[62] Hilborn DA. Serum stimulation of phosphate uptake into 3T3 cells. Journal of cellular physiology. 1976;87(1):111-121. DOI: 10.1002/jcp.1040870114

[63] Rubin H, Sanui H. Complexes of inorganic pyrophosphate, orthophosphate, and calcium as stimulants of 3T3 cell multiplication. Proceedings of the National Academy of Sciences USA. 1977;74(11):5026-5030. DOI: 10.1073/pnas.74.11.5026

[64] Abotaleb M, Kubatka P, Caprnda M, Varghese E, Zolakova B, Zubor P, Opatrilova R, Kruzliak P, Stefanicka P, Büsselberg D. Chemotherapeutic agents for the treatment of metastatic breast cancer: An update. Biomedicine & Pharmacotherapy. 2018;101:458-477. DOI: 10.1016/j.biopha.2018.02.108

[65] Lim EJ, Hong DY, Park JH, Joung YH, Darvin P, Kim SY, Na YM, Hwang TS, Ye SK, Moon ES, Cho BW. Methylsulfonylmethane suppresses breast cancer growth by down-regulating STAT3 and STAT5b pathways. PloS one. 2012;7(4):e33361. DOI: 10.1371/journal.pone.0033361

[66] Wang Z. Roles of K+ channels in regulating tumour cell proliferation and apoptosis. Pflügers Archiv. 2004;448(3):274-286. DOI: 10.1007/s00424-004-1258-5

[67] Frajese GV, Benvenuto M, Fantini M, Ambrosin E, Sacchetti P, Masuelli L, Giganti MG, Modesti A, Bei R. Potassium increases the antitumor effects of ascorbic acid in breast cancer cell lines in vitro. Oncology letters. 2016;11(6):4224-4234. DOI: 10.3892/ol.2016.4506

[68] Chen JQ, Contreras RG, Wang R, Fernandez SV, Shoshani L, Russo IH, Cereijido M, Russo J. Sodium/potasium ATPase (Na+, K+-ATPase) and ouabain/related cardiac glycosides: a new paradigm for development of anti-breast cancer drugs?. Breast cancer research and treatment. 2006; 96(1):1-15. DOI: 10.1007/s10549-005-9053-3

[69] Robey IF, Baggett BK, Kirkpatrick ND, Roe DJ, Dosescu J, Sloane BF, Hashim AI, Morse DL, Raghunand N, Gatenby RA, Gillies RJ. Bicarbonate increases tumor pH and inhibits spontaneous metastases. Cancer research. 2009;69(6):2260-2268. DOI: 10.1158/0008-5472.CAN-07-5575

[70] Anastassopoulou J, Theophanides T. Magnesium–DNA interactions and the possible relation of magnesium to carcinogenesis. Irradiation and free radicals. Critical reviews in oncology/hematology. 2002;42(1):79-91. DOI: 10.1016/S1040-8428(02)00006-9

[71] Wolf FI, Cittadini AR, Maier JA. Magnesium and tumors: ally or foe?. Cancer treatment reviews. 2009;35(4):378-382. DOI: 10.1016/j.ctrv.2009.01.003

[72] Mendes PM, Bezerra DL, dos Santos LR, de Oliveira Santos R, de Sousa Melo SR, Morais JB, Severo JS, Vieira SC, do Nascimento Marreiro D.

Magnesium in breast Cancer: what is its influence on the progression of this disease?. Biological trace element research. 2018;184(2):334-339. DOI: 10.1007/s12011-017-1207-8

[73] Huang WQ, Long WQ, Mo XF, Zhang NQ, Luo H, Lin FY, Huang J, Zhang CX. Direct and indirect associations between dietary magnesium intake and breast cancer risk. Scientific Reports. 2019;9(1):1-10. DOI: 10.1038/s41598-019-42282-y

[74] Atoe K, Idemudia O, Eboreime O. Serum magnesium levels in women with breast cancer in Benin City, Nigeria. International journal of tropical diseases and health. 2014;4(6):723-728. DOI: 10.9734/IJTDH/2014/5041

[75] Sahmoun AE, Singh BB. Does a higher ratio of serum calcium to magnesium increase the risk for postmenopausal breast cancer?. Medical hypotheses. 2010;75(3):315-318. DOI: 10.1016/j.mehy.2010.02.037

[76] Schrauzer GN. Anticarcinogenic effects of selenium. Cellular and Molecular Life Sciences CMLS. 2000 Dec 1;57(13-14):1864-1873. DOI: 10.1007/PL00000668

[77] McConnell KP, Jager RM, Bland KI, Blotcky AJ. The relationship of dietary selenium and breast cancer. Journal of Surgical Oncology. 1980;15(1):67-70. DOI: 10.1002/jso.2930150111.

[78] Li Z, Carrier L, Belame A, Thiyagarajah A, Salvo VA, Burow ME, Rowan BG. Combination of methylselenocysteine with tamoxifen inhibits MCF-7 breast cancer xenografts in nude mice through elevated apoptosis and reduced angiogenesis. Breast cancer research and treatment. 2009;118(1):33-43. DOI: 10.1007/s10549-008-0216-x

[79] Ibáñez E, Plano D, Font M, Calvo A, Prior C, Palop JA, Sanmartín C. Synthesis and antiproliferative activity of novel symmetrical alkylthio-and alkylseleno-imidocarbamates. European journal of medicinal chemistry. 2011;1;46(1):265-74. DOI: 10.1016/j.ejmech.2010.11.013

[80] Combs Jr GF, Clark LC, Turnbull BW. An analysis of cancer prevention by selenium. Biofactors. 2001;14(1-4):153-159. DOI: 10.1002/biof.5520140120

[81] Ruiz-Ramos R, Lopez-Carrillo L, Rios-Perez AD, De Vizcaya-Ruíz A, Cebrian ME. Sodium arsenite induces ROS generation, DNA oxidative damage, HO-1 and c-Myc proteins, NF-κB activation and cell proliferation in human breast cancer MCF-7 cells. Mutation Research/Genetic Toxicology and Environmental Mutagenesis. 2009;674(1-2):109-115. DOI: 10.1016/j.mrgentox.2008.09.021

[82] Wang Y, Zhang Y, Yang L, Cai B, Li J, Zhou Y, Yin L, Yang L, Yang B, Lu Y. Arsenic trioxide induces the apoptosis of human breast cancer MCF-7 cells through activation of caspase-3 and inhibition of HERG channels. Experimental and therapeutic medicine. 2011;2(3):481-486. DOI: 10.3892/etm.2011.224

[83] Fuster D, Herranz R, Vidal-Sicart S, Munoz M, Conill C, Mateos JJ, Martin F, Pons F. Usefulness of strontium-89 for bone pain palliation in metastatic breast cancer patients. Nuclear medicine communications. 2000;21(7):623-626.

Chapter 3

Analysis of Occurrence of Elements in Tissues of the Knee Joint

Wojciech Roczniak, Magdalena Babuśka Roczniak, Elżbieta Cipora and Barbara Brodziak Dopierała

Abstract

The mineral structure of bones is never static, it is a living structure, reacting and adapting to load and having the ability to remodel. Skeletal cells work continuously to maintain the remodelling process therefore they are in a constant state of dynamic balance both in the sense of composition and structure, and they react to external mechanical forces. The remodelling processes that occur in the bone tissue allow for a proper functioning of this tissue, as well as for inclusion of additional elements, toxic ones included, in the remodelled bone, and they affect the metabolic processes occurring therein. This may result in disturbances in the osteoarticular system, manifested by changes in the bone tissue and within other organs. The influence of tobacco smoking on the content of strontium, lead, calcium, phosphorus, sodium and magnesium has not been confirmed. Non-smokers showed a high iron content in knee joint tissues compared to smokers. There were no statistically significant differences in the content of cadmium, nickel, copper and zinc in women and men in the studied knee joint components. With age, an increase in the content of chromium in knee joint tissues was observed, while gender, place of residence and occupational exposure had no effect.

Keywords: knee joint tissues, structural and trace elements, environmental hazards

1. Introduction

Many joints can be distinguished in the human body, but one of them stands out in terms of function and size. It is the knee joint [1]. It belongs to a group of complex joints and connects the femur and the tibia together. This largest joint, apart from the mentioned elements, is formed by the sesamoid bone in the form of the kneecap, and two pieces of meniscus, which allow to match the joint surfaces to each other during movement. The knee joint allows making straightening and flexion movements, but also rotational movements possible only in incomplete joint flexion [2]. The entire structure of the knee joint is strengthened by strong internal and external ligaments. The afore-mentioned joint is the second most strained joint in the human body, after the ankle joint. Due to the powerful force that the quadriceps exerts on the kneecap (max. 300 kg), the knee joint is exposed to overload. Taking into account the functions of the knee joint, it must be both mobile and flexible, as well as resistant to pressure [3, 4].

The mineral structure of bones is never static, it is a living structure, reacting and adapting to load and having the ability to remodel. Skeletal cells work continuously to maintain the remodelling process therefore they are in a constant state of dynamic balance both in the sense of composition and structure, and they react to external mechanical forces. The remodelling processes that occur in the bone tissue allow for a proper functioning of this tissue, as well as for inclusion of additional elements, toxic ones included, in the remodelled bone, and they affect the metabolic processes occurring therein. This may result in disturbances in the osteoarticular system, manifested by changes in the bone tissue and within other organs. Maintaining all the characteristics of the knee joint is possible thanks to the balance of many elements of the bone tissue that are responsible for individual properties of bones, which in turn create separate joints. Elements occurring in large quantities, e.g. calcium, magnesium, phosphorus, and those with low content - the so-called trace elements, e.g. Strontium, can be found in the bone tissue. Regardless of the amount of elements contained in the bone tissue, all of them are important and play significant roles. Mostly, calcium and phosphorus are part of bone hydroxyapatite. However, during the mineralisation process, metal ions present in the blood plasma may be built into the bone tissue, and their uptake will depend on the affinity of a given metal for mineral and extracellular matrix, as well as the concentration of metal ions in the blood plasma, and the degree of skeletal mineralisation. Strontium present in the knee joint is a trace element, although it plays a special role in the bone remodelling process of the human body [5, 6]. It is accumulated mainly in bones due to the high similarity to calcium [7] but unlike it, it is absorbed from food much less efficiently and in a larger percentage it is excreted [8]. Previous in vivo studies demonstrate the effect of strontium on improvement of mechanical characteristics of bones [7]. These studies also proved the effectiveness of treatment with small doses of strontium in the form of strontium chloride. Their conclusion is that 9–26 week Sr. therapy activates bone building and also stops bone resorption in humans [8].

Iron is a cofactor in many enzymes and cells in redox reactions. Low levels of iron ions can be detrimental to cells, while an excess of iron ions can lead to the production of reactive oxygen species through the Fenton reaction. The cellular iron content is strictly regulated by homeostatic mechanisms to maintain the right amount of iron in cells. Nickel ions and other divalent metals can compete with iron ions to enter the cell through DMT1 (divalent metal transporter 1) because they have similar ionic radii. Therefore, metal ions can affect many other processes dependent on the presence of iron in cells. As an enzymatic cofactor, iron is involved in bone matrix synthesis (activation of lysyl hydroxylase) and in 25-hydroxy-cholecalciferol hydroxylase synthesis. What is more, thanks to active vitamin D, iron ions stimulate the absorption of calcium ions in the intestine. Iron deficiencies in rats led to poor mineralisation of skeletons and pathological changes in the micro-architecture of the spongy substance. In turn, administration of estrogens' increases the accumulation of iron in hamsters and facilitates the uptake of iron ions by lymphocytes in culture. The deficiency of iron ions in young rats leads to a decrease in the mechanical strength of femurs and the cortical bone. In severe iron deficiency, both bone strength and mineral density decrease. Excessive iron ion content in mice leads to increased oxidative stress. Oxidative stress mediates in bone loss through changes in bone remodelling. In rats with severe anaemia due to iron deficiency, the concentration of the N-terminal pro-collagen type I was low, which reduced bone formation and mineralisation [9]. These parameters returned to normal values after a diet with a normal iron content. There is no data, given the importance of iron for bone health in humans. Whereas, osteopenia was observed in patients with genetically determined hemochromatosis, and a very high iron content in tissues. Thus, the protective or destructive effects of iron on bones depend on its concentration. Iron deficiency related anaemia is still important for public health. Deficiencies of iron ions in women of childbearing age and in adolescents may also have an effect on bone health at a time when peak bone mass is achieved [9, 10].

Chromium in living organisms occurs as a trace element, yet its presence is extremely important. According to research on osteoblasts, chromium suppresses the level of osteocalcin the too high levels of which may accompany the osteoporosis process. It does not inhibit collagen production [11]. Reduced bone resorption in postmenopausal women has also been observed, and thus the prevention of osteoporosis together with an adequate level of chromium replenishment [12]. Due to its properties, chrome has been used in the production of orthopaedic implants. There are still concerns about local toxicity of chromium contained in prostheses, many studies address this issue [13–15]. Trace elements have a significant effect on the growth, development and condition of bone tissues. Changes in the mineral composition of bone tissues may cause degenerative changes and fractures. According to recent epidemiological data, the incidence of osteoarthritis around the world varies between 2 and 15% of the population. In Poland, this disease affects approximately 7–8 million people; in 40% of cases, degenerative changes are located within the hip joint and in 25% in the knee joint [8]. Trace elements have a significant effect on the growth, development and condition of bone tissues. Changes in the mineral composition of bone tissues may cause degenerative changes and fractures. The deficiency of some trace elements such as zinc, selenium, copper may increase the risk of bone resorption, inhibiting bone growth.

Environmental exposure to lead and cadmium is associated with the risk of a number of chronic diseases related to ageing, cardiovascular diseases, chronic renal failure and osteoporosis. The deficiency of some trace elements such as zinc, selenium, copper may increase the risk of bone resorption, inhibiting bone growth. Among metals that can affect the skeleton, there is no significant distinction between cellular effects and effects caused by the accumulation in the mineral or extracellular matrix. A given metal will undergo significant accumulation in the mineral and cause changes in its properties. Therefore, by studying specific mechanisms of the accumulation of metal ions, more information about bone mineralisation processes can be obtained. From a practical point of view, the influence of metal ions is often studied in relation to applied implants and prostheses.

2. Research goal

The goal of the manuscript was to determine the content of trace elements (Cd, Ni, Fe, Cr, Sr., Cu, Zn, Pb) and structural elements (Ca, P, Mg) in knee joint tissues. A diversified content of elements was determined in particular elements of the knee joint: tibia, femur and meniscus. Differences in the content of selected elements in the studied tissues between particular groups: women and men, smokers and non-smokers, inhabitants of cities and villages, people at risk of exposure and not exposed, patients operated on due to degenerative changes, and depending on age were considered in the research. The next stage of the research was a correlation analysis in the occurrence of elements, taking into account antagonistic and synergistic changes. For this purpose, various statistical methods were used to determine the dependence between the content of elements in bone tissues, e.g. main factors analysis and group similarity analysis.

Tissues that were examined were acquired intraoperatively during knee arthroplasty procedures based on the consent of the Bioethical Commission 2/2013 of 18.06.2013. The studied population consisted of women (n = 36) and men (n = 14) from 41 up to 82 years of age. Those people lived mainly in the areas of southern Poland, with the largest number of people coming from Upper Silesia.

3. Discussion of research results

Bone tissue has the ability to accumulate chemical elements and incorporate them into its structure, which is why it is often used to determine the impact of not only environmental but also occupational exposure.

Some metals such as zinc, iron and copper are closely related to human health because they are essential for maintaining normal physiological functions. However, heavy metal ions that are environmental pollutants show adverse health effects. Cadmium and lead can replace other elements that change the course of a number of biochemical reactions and can act as inhibitors, usually due to formation of complex compounds with sulphhydryl groups of proteins. Exposure to heavy metal compounds can affect genetic material and increase susceptibility to diseases. The World Health Organisation (WHO) classified some heavy metals such as cadmium, lead, mercury and arsenic as pollutants that need to be closely monitored [16]. The accumulation of an adequate amount of harmful heavy metal compounds in the human body changes the hormonal metabolism and narrows blood vessels. Metals are considered a risk factor for fractures and degenerative diseases in osteoporosis [17, 18].

As results from tests for the presence of elements, the average content of strontium in the entire knee joint reaches 17,50 mg/kg. There are no significant differences between Sr. depending on gender. The following strontium content can be distinguished in individual elements of the knee joint: meniscus - 1,44 mg/kg; femur - 24,60 mg/kg; while in tibia - 26,64 mg/kg. It is easy to see that the highest level of strontium in the examined knee joint is in the tibia, the lowest in the meniscus. The effect of smoking on the level of the element determined in the knee joint was also examined. The obtained results confirm a high level of Sr. in smokers compared to non-smokers, however the differences shown are not statistically significant.

Phosphorus present in the knee joint is the main component of all tissues of the human body. It plays a key role in mineralisation of the skeleton [19]. The content of phosphorus in individual bones is different. For example, in the femoral and tibial bones, the level of this element is 24 times higher than in the meniscus. The average content of phosphorus in the knee joint is 36,04 mg/kg. The obtained result is almost twice higher compared to strontium discussed above [20]. The level of phosphorus determined in the knee joint is slightly predominant among men, depending on gender groups. The differences generated in the study did not reach the statistically significant level.

In the case of Pb, Ca, P, Na, Mg content, significant statistical differences occurred in individual elements of the knee joint (Kruskal-Wallis ANOVA test, $p < 0{,}001$). The lead content in the meniscus was 0,32 mg/g, in the tibia 2,67 mg/g, and in the femur 2,64 mg/g. The highest calcium content was in the tibia – 122,57 g/kg, and in the femur – 112,45 g/kg, in the meniscus the content was about 23 times lower and was 5,08 g/kg. The phosphorus content was similar, the highest in the tibia – 55,34 g/kg and in the femur – 50,56 g/kg, and the lowest in the meniscus – 2,21 g/kg. In the case of sodium in the tibial and femoral bones, the content was 5,50 and 5,56, and in the meniscus – 2,11 g/kg. The magnesium content was as follows: tibia 1,55, femur 1,42, and meniscus 0,10 mg/kg. The highest content of Sr., Pb, Ca, P, Na, Mg was in the tibia and the smallest in the meniscus. Statistically significant differences between men and women occurred only in the tibia and related to lead content (U Mann–Whitney U test, $p = 0{,}011$).

The influence of tobacco smoking on the content of strontium, lead, calcium, phosphorus, sodium and magnesium did not cause statistically significant differences. Among those elements, only the content of strontium was greater in people who smoked tobacco. Whereas, the contents of lead, calcium, phosphorus, sodium and magnesium were higher in non-smokers. There were no differences between individual elements of the knee joint and sex and smoking.

Another element that is significant and present in the knee joint is iron. As an enzyme cofactor, it participates in the formation of bone matrix. Iron deficiency in the human body, especially a significant deficiency, leads to a decrease in both density and bone strength. Deficiency, as well as excess of Fe adversely affects the system. Too high value of the element increases oxidative stress [21]. According to research on rats, iron deficiency leads to diseased changes in the spongy substance of discs and weakened skeletal mineralisation. For this reason, the mechanical strength of the femoral bones decreases considerably [22]. The effect of iron content on the health of human bones has not yet been studied. There are works on Fe that suggest that maintaining high levels of iron may help in the prevention of bone fractures in older women [21]. Studies concerning iron content in selected knee joint tissues are slightly different when compared to other elements studied. In the case of iron, its highest content was found in the femur, in which the level is 41,91 μg/g. The second, extreme iron value was determined at 27,04 μg/g in the tibial bone. An intermediate level was determined in the meniscus - i.e. 38,68 μg/g. There were no statistically significant differences between the tibial and the femoral bones in gender groups, but the marked values among women predominated over those marked in the opposite sex. The marked Fe content in the meniscus was different in the group of men. Differences turned out to be statistically insignificant in this case as well. The studies also included iron levels in the knee joint tissues in relation to smokers and non-smokers. In non-smokers, high Fe levels (39,11 μg/g) were determined compared to smokers (25,47 μg/g). The last examined dependence related to the iron level in knee joint tissues was the operation of knee arthroplasty. The study of this correlation shows that Fe levels are lower in patients with a knee prosthesis implanted (32,81 μg/g < 36,96 μg/g).

Chromium is found in living organisms as a trace element. Its presence is extremely important, it inhibits the level of osteocalcin the high level of which accompanies the osteoporosis process, as results from studies on osteoblasts. Reduced bone resorption in postmenopausal women has also been observed, and thus the prevention of osteoporosis together with an adequate level of chromium replenishment [23]. Thanks to its properties, Cr has been used in the production of orthopaedic implants. Nevertheless, there are many concerns about possible local toxicity of prostheses with its content. This problem has been addressed in many manuscripts [13]. As in the case of iron, the highest chromium content in the knee joint occurs in the femur (1,64 μg/g). Slightly lower levels are found in the tibial bone (1,27 μg/g), with the lowest in the meniscus (1,18 μg/g). A slightly higher level of Cr is recorded among men however without statistical significance. An interesting examined dependence is the increase in chromium levels in the knee joint along with age. Among the respondents, its highest level (1,78 μg/g) was recorded in people over 70 years of age. The Cr content of the knee joint in residents of cities is almost twice as high (2,30 μg/g) compared to inhabitants of villages (1,20 μg/g). In smokers, a higher level of some metals in the body can be seen which is due to their presence in tobacco smoke. In the case of chromium, this dependence was not confirmed. Its lower level was examined in people smoking cigarettes (1,00 μg/g) compared to non-smokers (1,47 μg/g). The content of elements in the knee joint tissues in female and male were showed in **Figure 1**.

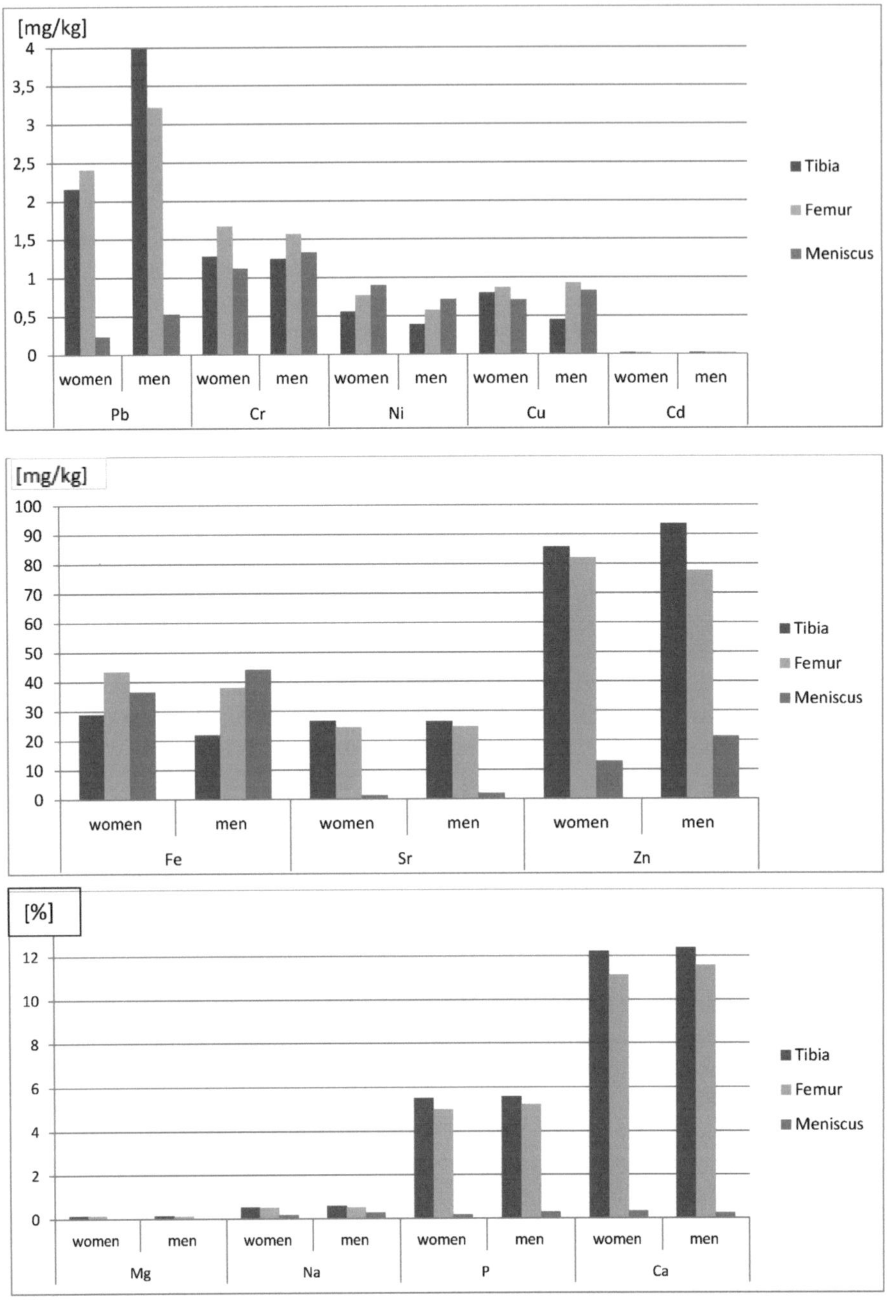

Figure 1.
The content of elements in the knee joint tissues in female and male.

Apart from the elements discussed above, the presence of nickel, cadmium, zinc and copper in the knee joint can be distinguished. Many factors affect their level. These include: the type of tissue being examined, gender, place of residence, nicotinism, age, occupational exposure. The lowest content in knee joint tissues is shown by cadmium. Nickel is characterised by a higher level in women's knee joints (tibia – 0,29 μg/g, femur – 0,36 μg/g, meniscus – 0,69 μg/g) in relation to men (tibia – 0,22 μg/g, femur – 0,28 μg/g, meniscus – 0,42 μg/g). The lowest percentage of copper in the knee joint in women concerns the femur (0,36 μg/g), and in men the tibial bone (0,31 μg/g). In the studied joint, the element that has an advantage over the others is zinc. In smokers, an increased level of cadmium in knee joint tissues is observed due to its content in tobacco smoke. Zinc affects the condition, growth and development of bone tissue. The deficiency of zinc or copper leads to an increased bone resorption and thus inhibits their growth. What is more, zinc is responsible for the activity of vitamin D. Its deficiency leads to osteoporosis.

Keeping the balance in the mineral composition of bone tissue is very important and any deviations from the norm may cause degenerative changes and fractures [24].

4. Conclusions

The results of the presented research results indicate that the bone tissue of the femur and the tibia of the knee joint can be used to determine the content of such elements as lead, cadmium, chromium, zinc, magnesium, potassium and calcium. There was 24 times more phosphorus, 23 times more calcium, 18 times more strontium, 15 times magnesium, 8 times lead, and 3 times sodium in the femur and the tibia compared to the meniscus. However, copper and nickel showed a high content in connective tissue (meniscus) compared to bone tissue (tibia and femur). High values of metals can affect the structure of bone tissue and cause a change in composition and its properties. One of the most common correlations described in the literature on the subject has been confirmed - it is a synergistic correlation between nickel and copper.

The influence of tobacco smoking on the content of strontium, lead, calcium, phosphorus, sodium and magnesium has not been confirmed. Non-smokers showed a high iron content in knee joint tissues compared to smokers. There were no statistically significant differences in the content of cadmium, nickel, copper and zinc in women and men in the studied knee joint components.

With age, an increase in the content of chromium in knee joint tissues was observed, while gender, place of residence and occupational exposure had no effect.

Author details

Wojciech Roczniak[1]*, Magdalena Babuśka Roczniak[1], Elżbieta Cipora[1] and Barbara Brodziak Dopierała[2]

1 The Jan Grodek Higher Vocational State School, Medical Institute, Sanok, Poland

2 Department of Toxicology and Bioanalysis, School of Pharmacy with the Division of Laboratory Medicine, Medical University of Silesia, Sosnowiec, Poland

*Address all correspondence to: wojciech_roczniak@interia.pl

References

[1] Kapandji A.I.: Anatomia funkcjonalna stawów. Kończyna dolna TOM 2. Elsevier Urban & Partner, Wrocław 2013.

[2] Nordin M., Frankel V.H.: Basic biomechanics of the musculoskeletal system. 4th edition. Wolter Kluwer; Lippincott Wiliams & Wilkins 2012.

[3] Lemiesz G.I.: The effectiveness of rehabilitation procedure after the reconstruction of the anterior cruciate ligament according to the Norwegian protocol. Pol. Ann. Med. 2011; 18(1): 82-95.

[4] Trzaska T.: Aktualne metody rekonstrukcji więzadła krzyżowego przedniego. Medicina Sportiva 2012; 6: 19-22.

[5] Brodziak-Dopierała, B.; Kwapuliński, J.; Kusz, D.; Gajda, Z.; Sobczyk, K.: Interactions between concentrations of chemical elements in human femoral heads. Arch. Environ. Contam. Toxicol. 2009; 57: 203-210.

[6] Branca, F.; Vatuena, S.: Calcium, physical activity and bone health—Building bones for stronger future. Public Health Nutr. 2001; 4: 117-123.

[7] Nordberg, G.F.; Fowler, B.A.; Nordberg, M.; Friberg, L.T.: Handbook on the Toxicology of Metals. Elsevier: London, UK, 2008.

[8] Uenishi, K.: Nutrition and bone health. Conclus. Clin. Calcium 2010; 20: 940-943.

[9] Diaz-Castro, J.; Lopez-Frias, M.R.; Campos, M.S.; Lopez-Frias, M.; Alferez, M.J.M.; Nestares, T.; Ojeda, M.L.; Lopez-Aliaga, I.: Severe nutritional iron-deficiency anemia has a negative effect on some bone turnover biomarkers in rats. Eur. J. Nutr. 2012; 51: 241-247.

[10] Balogh, E.; Paragh, G.; Jeney V. Influence of iron on bone homeostasis. Pharmaceuticals 2018; 11(4): 107.

[11] Allen, M.J.; Myer, B.J.; Millett, P.J.; Rushton, N.: The effects of particulate cobalt, chromium and cobalt-chromium alloy on human osteoblast-like cells in vitro. J. Bone Jt. Surg. Br. 1997; 79: 475-482.

[12] McCarty, M.F.: Anabolic effects of insulin on bone suggest a role for chromium picolinate in preservation of bone density. Med. Hypotheses 1995; 45: 241-246.

[13] Scharf, B.; Clement, C.C.; Zolla, V.; Perino, G.; Yan, B.; Elci, S.G.; Purdue, E.; Goldring, S.;Macaluso, F.; Cobelli, N.; Vachet, R.W.; Santambrogio, L.: Molecular analysis of chromium and cobalt-related toxicity. Sci. Rep. 2014; 4: 5729.

[14] Wang, J.Y.; Wicklund, B.H.; Gustilo, R.B.; Tsukayama, D.T.: Titanium, chromium and cobalt ions modulate the release of bone-associated cytokines by human monocytes/ macrophages in vitro. Biomaterials 1996; 17: 2233-2240.

[15] Rakow, A.; Schoon, J.; Dienelt, A.; John, T.; Textor, M.; Duda, G.; Perka, C.; Schulze, F.; Ode, A.: Influence of particulate and dissociated metal-on-metal hip endoprosthesis wear on mesenchymal stromal cells in vivo and in vitro. Biomaterials 2016; 98: 31-40.

[16] Rebolledo, J.; Fierens, S.; Versporten, A.: Brits, E.;, De Plaen, P.; Nieuwenhuyse, A.V.: Human biomonitoring on heavy metals in Ath: methodological aspects. Arch. Public Health 2011; 69: 10.

[17] Kido, S.: Secondary osteoporosis or secondary contributors to bone loss in fracture. Bone metabolism and heavy

metals (cadmium and iron). Clin. Calcium 2013; 23: 1299-1306.

[18] Hee-Sook, L., Hae-Hyeog, L., Tae-Hee, K., Bo-Ra L.: Relationship between heavy metal exposure and bone mineral density in Korean adult. J. Bone Metab. 2016; 23(4): 223-231.

[19] Carter, D.R.; Beaupré, G.S.: Skeletal function and form: mechanobiology of skeletal development, aging and regeneration. Cambridge University Press: Cambridge, UK, 2007.

[20] Dahl, S.G.; Allain, P.; Marie, P.J.; Mauras, Y.; Boivin, G.; Ammann, P.; Tsouderos, Y.; Delmas, P.D.; Christiansen, C.: Incorporation and distribution of strontium in bone. Bone. 2001; 28: 446-453.

[21] Zofková. I.; Nemcikova, P., Matucha, P.: Trace elements and bone health. Clin. Chem. Lab. Med. 2013; 51: 1555-1561.

[22] Parelman, M.; Stoecker, B.; Baker, A, Medeiros *D. iron* restriction negatively affects bone in female rats and mineralization of hFOB osteoblast cells. Exp. Biol. Med. 2006; 231: 378-386.

[23] Brodziak-Dopierała B., Roczniak W., Jakóbik-Kolon A., Kluczka J., Koczy B., Kwapuliński J., Babuśka-Roczniak M.: Correlations between iron kontent in knee joint tissues and chosen indices of peripheral blond morphology. Adv. Clin. Exp. Med. 2017; 26 (7): 1077-1083.

[24] Brodziak-Dopierała, B., Kwapuliński, J., Sobczyk K., Wiechuła, D: Distribution of magnesium, calcium, sodium and potassium in tissues of the hip joint. Magnes. Res. 2013; 26: 125-131.

Chapter 4

Natural Iron Chelators as Potential Therapeutic Agents for the Treatment of Iron Overload Diseases

Naheed Waseem A. Sheikh, Satish B. Kosalge, Tusharbindu R. Desai, Anil P. Dewani, Deepak S. Mohale and Alok S. Tripathi

Abstract

Iron overload disease is a group of heterogeneous disease, which is caused either due to hereditary or acquired condition. Excess of iron participate in redox reactions that catalyzes the generation of reactive oxygen species (ROS) and increases oxidative stress, which causes cellular damage and encourage the cell injury and cell death. The electronic databases of Scopus, PubMed and Google Scholar have been intensively searched for the research as well as review articles published with the full text available and with the key words such as natural iron chelating agent, synthetic iron chelating agents, iron overload disease, oxidative stress and antioxidant which were appearing in the title, abstract or keywords. In light of the literature review presented in this artial, based on meta-analyses, we suggest that iron chelating agents were used for the management of iron overload disease. These agents were having wide spectrum of activity, they were not only used for the management of iron overload disease but also used as anticancer and antioxidant in various oxidative stress mediated diseases. Last from many years Desferoxamine (DFO) was used as standard iron chelator but currently two new synthetic iron chelators such as Deferiprone (DFP) and Deferasirox (DFS) are available clinically. These clinically available synthetic iron chelators were having serious side effects and certain limitations. Phytochemicals such as flavonoids and polyphenols compounds were having iron chelating as well as antioxidant property with no or minimal side effects. Hence, this review provides an updates on natural iron chelation therapy for the safe and efficacious management of iron overload diseases.

Keywords: Natural iron chelating agents, synthetic iron chelating agents, iron overload disease, oxidative stress, antioxidant

1. Introduction

Iron is an essential element for all living organism. It participates in various biochemical reactions like oxygen transport, electron transfer, energy metabolism and DNA synthesis. The biological action of iron is largely due to its chemical properties as a transition metal, although these properties make it potentially toxic. Total body iron content of an adult is 3–5 g (~ 55 mg/kg for male and ~ 45 mg/kg female). Majority of body iron is incorporated within hemoglobin (Hb) (60–70%) as circulating RBC; while about 20–30% of iron is present in the form of ferritin and hemosiderin as a spare iron in hepatocytes and reticuloendothelial (RE) macrophages. Remaining body iron is primarily located in myoglobin and enzymes such as cytochromes, peroxidases, catalases, xanthine oxidase and some mitochondrial enzymes. A healthy adult absorbs near about 1–2 mg/day of iron from the diet, which reimburses the non-specific loss of iron by exfoliation of intestinal epithelial cells. Moreover menstruating women additionally losses iron during the menstrual cycle. Erythropoiesis daily requires about 30 mg of iron, which is largely provided by the recycling of iron through RE macrophages.

The dietary iron is absorbed from the small intestine and the normal level of iron is regulated by feedback mechanism between its requirement and absorption. The dietary iron is present either as haeme or as inorganic ferric iron ($Fe3+$). Absorption of haeme iron takes place directly without the aid of active carrier transport. However, haeme iron is a minor source of dietary iron. The primary source of dietary iron is $Fe3+$ which has to be reduced to ferrous ($Fe2+$) form by acid reducing agents for efficient uptake. Iron is transported across the membrane by two distinct transporters. The divalent metal transporter 1 (DMT1) present at luminal membrane carries $Fe2+$ into the intestinal epithelial cell. This $Fe2+$ and iron released from the haeme is transported across the basolateral membrane by another iron transporter ferroportin (FP). Gut has a mechanism to prevent the entry of excess iron in the body. After reaching to the intestinal epithelial cell, iron is either transported to plasma or oxidized to $Fe3+$ and complexed with apoferritin to form ferritin, the cytosolic protein in which iron is stored. Ferritin usually remains stored in intestinal epithelial cells for 2–4 days after that the cells were shed off. This process is called as exfoliation. Whenever the body iron is low, the ferritin is either not formed or dissociates quickly to release iron. This released iron is transported to the blood [1].

The free form of iron is extremely toxic. The $Fe2+$ on entering into plasma it is rapidly oxidized to $Fe3+$ and complexed with apotransferrin (Apo-Tf) a glycoprotein to form transferrin (Tf). Two $Fe3+$ residues bound to one Tf molecule. This complex bound to membrane bound transferrin receptor 1 (Tfr1), present on erythropoietic and other cells. The Tf–Tfr1 complex is engulfed by receptor mediated endocytosis. The endosomes of erythropoietic and other cells become acidified through engulfed proton complex, which leads to conformational changes, which dissociates iron from the complex. The released $Fe3+$ is reduced to $Fe2+$ and transported out of the endosomes by DMT1. This released $Fe2+$ is utilized for hemoglobin synthesis or other biochemical process; Apo-Tf and Tfr1 are return to the cell surface for further cycles. In iron deficiency and haemolytic anemia, Tfr1 receptors up regulation take place at erythropoietic cells, but not at other cells. Under physiological conditions, all circulating iron is bound to transferrin. Nontransferrin bound iron (NTBI) can increase in a pathological condition like iron overload disease (**Figure 1**) [2, 3, 14].

Once the iron enters the cell, the fraction that is not needed for immediate use is stored by ferritin and haemosiderin in RE cells of liver, spleen and bone marrow. Iron status of the body regulates the synthesis of apoferritin. When the iron status is low, the iron regulating element (IRE) at DNA is blocked and synthesis of apoferritin is not taken place, whereas more Tf is produced. Moreover excess of apoferritin is synthesized to trap iron when the iron store is high [4, 5].

The plasma iron obtained from three primary sources, firstly from constant degradation of older RBC (lifespan ~120 days), secondly from stored iron from RE cells in liver, spleen and bone marrow while thirdly from intestinal absorption. The conservation and recycling of iron are necessary to reload the iron contained within Hb. The recycling takes place at macrophage, which phagocytes the RBC and liberates iron in haeme form by haeme oxygenase-1 [6].

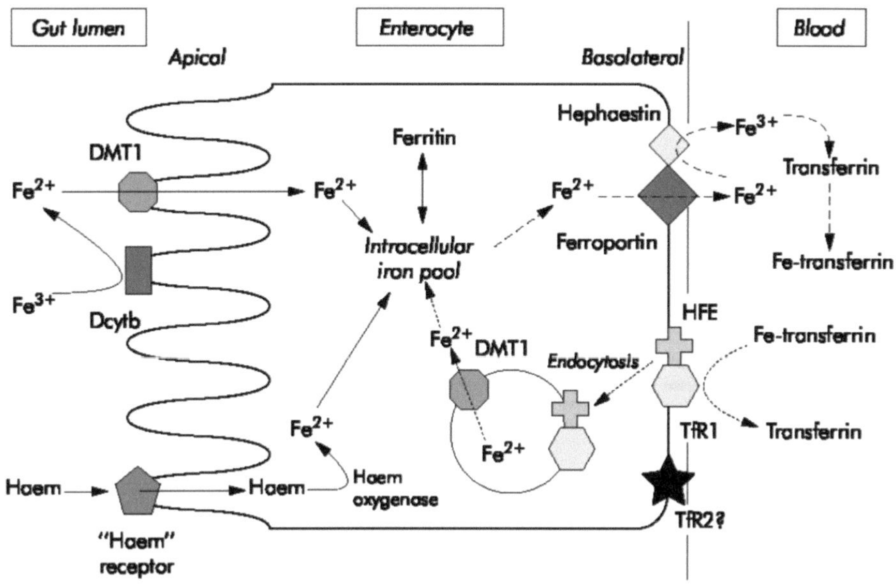

Figure 1.
Iron absorption pathways by the enterocyte.

Daily excretion of iron in an adult male is approximately 1–2 mg, mostly as exfoliated intestinal epithelial cells, some RBCs and in bile. The primary route of excretion of iron is feces, while skin, urine and sweat are minor routes. In menstruating women, monthly menstrual loss of iron is about 0.5–1 mg/day. Excess of iron is required during last two trimesters of pregnancy for expansion of RBC mass, transfer to the fetus and to compensate the loss during delivery [7].

The iron homeostasis maintained by its controlled absorption, recycling and storage. The iron is regulating peptide hormone, hepcidin produced mainly by the liver, whereas a smaller amount is produced in other organs like lung and heart plays a vital role in this regards. Hepcidin act by degradation of FP, an iron efflux transporter, concerned about transportation of iron across the basolateral membrane in the intestine. Iron overload increases whereas anemia and hypoxia decrease the synthesis of hepcidin [8]. Hepcidin synthesis is regulated by bone morphogenetic protein (BMP)/SMAD pathway via activation of hepcidin transcription. The loss of hepatic SMAD4 gene results in iron overload due to the failure of hepcidin-mediated degradation of FP [9].

2. Determination of serum iron

The biochemical estimation of body iron status depends on serum based indicators, as follows

1. Serum iron (SI)

2. Serum ferritin (SF)

3. Transferrin saturation

4. Soluble transferrin receptor (sTfR)

5. Erythrocyte protoporphyrin

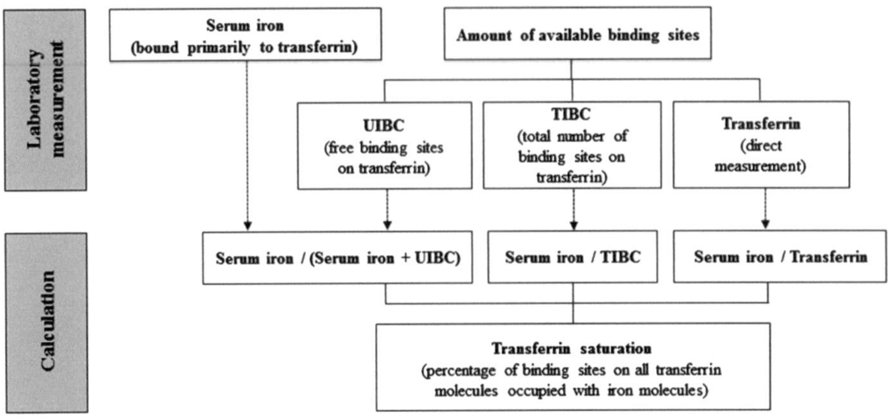

Figure 2.
Laboratory measurement of iron indicators needed to calculate transferrin saturation.

These indicators present challenges for clinical practice and national nutrition surveys, and often iron status interpretation is based on the combination of several indicators (**Figure 2**). The diagnosis of iron deficiency through SF concentration, the most commonly used indicator, is complicated by concomitant inflammation. The sTfR concentration is an indicator of functional iron deficiency that is not an acute phase reactant, but challenges in its interpretation arise because of the lack of assay standardization, common reference ranges, and common cutoffs. However it is unclear which indicators are best suited to assess excess iron status. The value of hepcidin, non–transferrin-bound iron, and reticulocyte indexes is being explored in research settings. Serum based indicators are generally measured on fully automated clinical analyzers. Although international reference materials have been available for years, the standardization of immunoassays is complicated by the heterogeneity of antibodies used and the absence of physicochemical reference methods to establish "true" concentrations. The assessment of iron status in NHANES was based on the multi-indicator ferritin model. However, the model did not indicate the severity of iron deficiency and produced categorical estimates. Recently, iron status assessment in NHANES has used the total body iron stores (TBI) model, in which the log ratio of sTfR to SF is assessed. Together, sTfR and SF concentrations cover the full range of iron status. The TBI model better predicts the absence of bone marrow iron than SF concentration alone, and TBI can be analyzed as a continuous variable. Additional consideration of methodologies, interpretation of indicators and analytic standardization is important for further improvements in iron status assessment [10].

3. Iron overload disease

Iron overload disease is also known as haemochromatosis is a group of heterogeneous disease which is caused either due to hereditary or acquired condition. Iron overload disease is characterized by the accumulation of iron in the body with or without organ dysfunction [11, 12]. Iron overload is unavoidable since there is no physiologically regulated mechanism for excretion of excess iron. During iron overload, the low molecular weight iron can play an essential catalytic role in the initiation of free radical reactions. These free radicals have the potential to damage cellular macromolecules like lipids, proteins, carbohydrates and nucleic acids resulting in cell injury, impaired cell function and integrity or cell death. The rate of free radical generation determines the intensity of cell injury [13].

3.1 Types of iron overload disease (haemochromatosis)

The haemochromatosis is classified into two main categories, namely primary and secondary haemochromatosis [14].

A. Familial or hereditary haemochromatosis (Primary iron overload).

   a. Hereditary haemochromatosis (HH, HFE1).

      i. C282Y homozygosity.

      ii. C282Y, H63D heterozygosity.

      iii. Other HFE gene mutations.

   a. Juvenile haemochromatosis (HFE2).

   b. Transferrin receptor 2 mutation (HFE3).

   c. Ferroportin mutation (HFE4).

   d. Aceruloplasminemia.

   e. Atransferrinaemia.

   f. Neonatal iron overload.

   g. Autosomal dominant haemochromatosis (Solomon Islands).

B. Acquired haemochromatosis (Secondary iron overload).

   a. Iron loading anemia's.

   b. Transfusional iron overload.

   c. African iron overload.

   d. Iron overload in chronic liver disease.

3.1.1 Familial or hereditary haemochromatosis (primary iron overload)

Hereditary haemochromatosis (HH, HFE1) is the most prevalent form of primary iron overload disease. HFE1 is an autosomal recessive disorder caused due to a mutation in HFE gene on chromosome 6, that resulting in iron overload and variable multiorgan dysfunction. More than 20 mutations of HFE gene were identified, but the most clinically significant mutations, however, are the C282Y and H63D. The C282Y mutation is a missense mutation that causes the tyrosine to replace cysteine at position 282, whereas H63D mutation is characterized by a histidine to aspartic acid substitution at position 63 in the HFE protein [15]. The H63D mutation may add to minor increases in serum iron levels, but in the absence of C282Y, there was no clinical significance of the H63D mutation. Approximately 85–90% of HFE patients were C282Y homozygotes while 3–5% of subjects with HFE may be C282Y/H63D compound heterozygous [16]. Another mutation of HFE gene is S65C that leads to substitution of serine for cysteine, this kind of mutation also results into mild iron overload in the compound to heterozygous type HFE1 [17].

The C282Y mutation leads to disruption of a disulfide bridge that decreases the affinity of HFE gene towards β2 microglobulin and TfR1. This type of HFE gene mutation is retained in the Golgi complex [18]. This mutation resulted in decreased uptake of plasma transferrin-bound iron (TBI) by the duodenal crypt cells. This decreased uptake of iron would result into false iron deficiency even increasing total body iron stores that results into upregulation of iron regulating protein (IRP), DMT1 and ferroportin 1 (Fpn1) expression and increased iron absorption from the intestine [19, 20].

Mutation of a gene other than HFE gene is a rare genetic disorder of iron metabolism, also responsible for the development of the iron overloaded disease. Juvenile haemochromatosis, a type 2 haemochromatosis (HFE2) is a rare autosomal recessive disorder resulting in iron overload during the second and third decades of life. Some patients with HFE2 have mutations in hepcidin or hemojuvelin genes [21, 22]. Mutations of transferrin receptor-2 gene situated on chromosome 7q22 responsible for the development of transferrin receptor 2 mutations (HFE3) type of haemochromatosis. HFE3 is inherited in an autosomal recessive fashion [23]. HFE4 is caused due to mutations in ferroportin gene (IREG1, MTP1 or SLC11A3). It is inherited in an autosomal dominant fashion. These mutations are rare and should not routinely be screened for diagnosis purpose, but should be considered if HH cannot be diagnosed by conventional HFE gene mutations [24]. Other forms of hereditary haemochromatosis like aceruloplasminemia, atransferrinaemia, neonatal iron overload and autosomal dominant haemochromatosis (Solomon Islands) are rare conditions and they comprise of the very negligible proportion of inherited haemochromatosis [25].

3.1.2 Acquired haemochromatosis (secondary iron overload)

The ineffective erythropoiesis like thalassaemia, hereditary sideroblastic anemia and certain myelodysplastic syndromes, the hyperplastic erythroid marrow stimulate the iron absorption to a level that leads to the clinical iron overloaded condition. There is a direct relationship between erythropoiesis and iron absorption [26]. Each unit of blood contains 200–250 mg of iron. Chronic blood transfusion therapy is required in a condition like β thalassaemia, bone marrow failure and sickle cell anemia. Excess of transfusion, the iron load, initially accumulates in the RE macrophages but iron may deposit later in the parenchymal cells of the liver, heart, pancreas and endocrine tissue. Phlebotomy is not a treatment alternative because of the underlying anemia. Iron chelation therapy with Desferoxamine (DFO) administered continuous infusion is often the only option. Another oral iron chelator, Deferiprone (DFP) and Deferasirox (DFS) has been studied but has limitations in their efficacy [27].

Iron overload in sub-Saharan Africa was believed initially due to ingestion of large amounts of iron obtained from traditional home brewed beer fermented in non-galvanized steel drums. Moreover, only a few numbers of these beer drinkers acquire iron overload, suggesting that a genetic predisposition may be involved in the development of iron overload. This indicates that the gene locus is not related to the HFE gene, but maybe specific putative locus has not been identified [28]. Whereas heterozygosity for a common polymorphism (Q248H) in Fpn1 gene was identified in African and African-American person with iron overload [29].

Liver cirrhosis may increase the hepatic iron deposition. It seen in non-biliary cirrhosis likes alcoholic liver disease, chronic viral hepatitis and non-alcoholic steatohepatitis [30]. Very few cases were reported to have iron overload due to liver cirrhosis. The mechanism is not known, heterozygosity for the C282Y mutation in the HFE gene may play an important role [31]. In cirrhosis, there is a decrease in the synthesis of Tf whereas an increase in the plasma NTBI levels, which may contribute to hepatic iron overload [32]. Reduced incorporation of iron into the RBC as well as reduced RBC lifespan in liver cirrhosis related hypersplenism, combined with portocaval shunting may also contribute to the liver iron overload [33].

3.2 Complications of iron overload disease

Iron overload may lead to produce organ complications such as liver failure (fibrosis and rarely carcinoma), Cardiac abnormalities (cardiomyopathy, arrhythmias and heart failure), and hepatosplenomegaly. The other symptoms include chronic abdominal pain, weakness, fatigue, joint pain, arthritis, osteoporosis, arthralgia, loss of libido, impotence, infertility, hyperpigmentation of the skin, cutaneous atrophy, flattening of nails and loss of hair [12]. Endocrine abnormalities like diabetes mellitus, hypothyroidism, hypoparathyroidism, hypocortisolism, adrenal insufficiency hypothalamic–pituitary dysfunction, pancreatic dysfunction, amenorrhoea, delayed puberty and hypogonadism seen with iron overloaded patients [34]. Iron overload also leads to kidney damage [35]. Accumulation of iron in the brain may lead to neurodegenerative diseases like Alzheimer disease, Parkinson disease, amyotrophic lateral sclerosis, multiple sclerosis, progressive supranuclear palsy, corticobasal degeneration and superficial siderosis [36].

4. Iron chelator: pharmacology and toxicology

4.1 Concept and chemistry of iron chelator

Iron chelators typically contain donor atoms like oxygen, nitrogen or sulfur which form coordinate bonds with the bound iron. The donor atoms determine the preference of the chelator for either the Fe2+ or Fe3+ oxidation states [37]. Chelators that contain nitrogen and sulfur as donor atoms can prefer not only Fe2+ but also other divalent metals such as Cu2+ and Zn2+ [38]. Iron chelators may be classified by their binding structures. Bidentate iron chelator such as DFP requires three molecules each with two iron binding sites Fe3+ (3:1 ratio). A tridentate iron chelator DFS requires two molecules for Fe3+ (2:1 ratio); whereas hexadentate iron chelator, DFO binds Fe3+ in a 1:1 ratio [39]. Iron can coordinate six ligands in an octahedral arrangement. Hence DFO has the highest affinity for iron. The effectiveness of iron chelator determine by how wholly and efficiently it form the complex; thus affinity and stoichiometry of iron chelator play an essential role for its therapeutic effectiveness [40].

Iron chelators were mainly focused on the management of iron overload conditions due to multiple blood transfusions as the supportive treatment of disease like β-thalassaemia, sickle cell disease and myelodysplasia [41]. An iron chelator is having a broad spectrum of activity, they were not only used for the management of iron overload disease, but also as in the treatment of cancer due to their ability to sequester metals essential to tumor growth [42]. Other than the iron chelation, they also play an essential role as antioxidant in various oxidative stress mediated diseases like liver disease [43], ischemic reperfusion injury [44], atherosclerosis [45], diabetes mellitus [46], inflammation [47], infectious disease [48] and neurologic disease [49].

4.2 Current iron chelator

In current medicine, iron chelators include natural compounds derived from microorganisms such as siderophores and synthetic iron chelators were clinically used for the treatment of iron overloaded conditions.

4.2.1 Siderophores

Siderophores are the low molecular mass with high affinity iron chelating compounds that are secreted by the iron dependent microorganisms such as bacteria and fungi. They serve primarily as iron transport across the cell membrane. Wide range of siderophores is available such as Ferrichrome, DFO, Fusarinine, Ornibactin, Enterobactin, Bacillibactin, vibriobactin and Azotobactin. The commonly used siderophore is DFO [50].

4.2.1.1 Desferoxamine (DFO)

DFO is the most common clinically used siderophore. It has been used for the treatment of iron overloaded diseases last for decades and it remains the current standard for the iron chelation therapy [51]. The toxic effect of an excess of iron in iron overloaded disease is majorly due to NTBI; iron has both redox activity as well as ability to concentrate in highly vascular tissues such as hepatic, cardiac and endocrine tissue [52]. DFO is a hexadentate iron chelator, which can bind Fe3+ in a 1:1 ratio as shown in **Figure 3**. DFO is a multifunctional therapeutic agent, which can detoxify NTBI by its chelation property and heme proteins by ferryl reduction as well as free radical scavenging action. DFO also acts as a reducing agent, which prevent the oxidation of membrane lipids by removing high-oxidation states of heme iron, like ferryl myoglobin (Mb) or Hb [53].

Because of its high molecular weight (656.79), DFO is not orally bioavailable. Hence it is administered via subcutaneous injection at a dose of 50 mg/kg/day as a 10% solution in sterile water (0.50 or 2.0 g vials). Additionally, DFO has a short half-life of about 5–10 min, therefore to improve its efficacy; the required dose is injected over a period of 4–12 hrs via a small portable peristaltic pump. DFO is poorly metabolized by transamination, β -oxidation, decarboxylation and N-hydroxylation. DFO is excreted as its 1:1 complex with iron mostly in the urine and a small amount in feces [54].

4.2.2 Synthetic iron chelator

4.2.2.1 Deferiprone (DFP)

DFP is a synthetic oral iron chelator that has shown comparable efficacy to DFO and is more effective than DFO in the removal of excess iron from the heart. An advantage of DFP is that Fe3+ chelate of DFP carries no net charge and therefore, DFO-iron complex can easily penetrate the membrane. Additionally, the combination of DFO and DFP is widely used now a day without any new toxic effects [51].

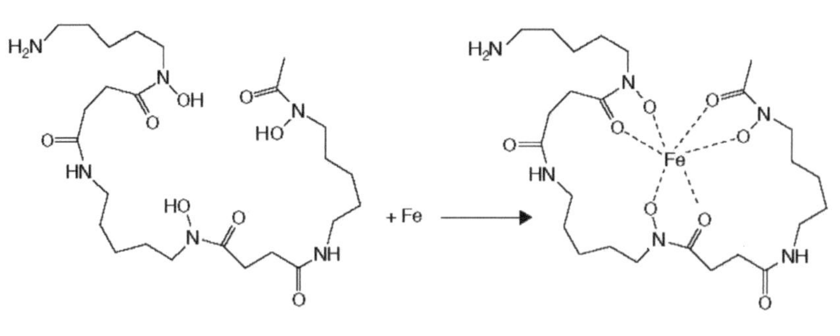

Figure 3.
Chelation of iron with Desferoxamine.

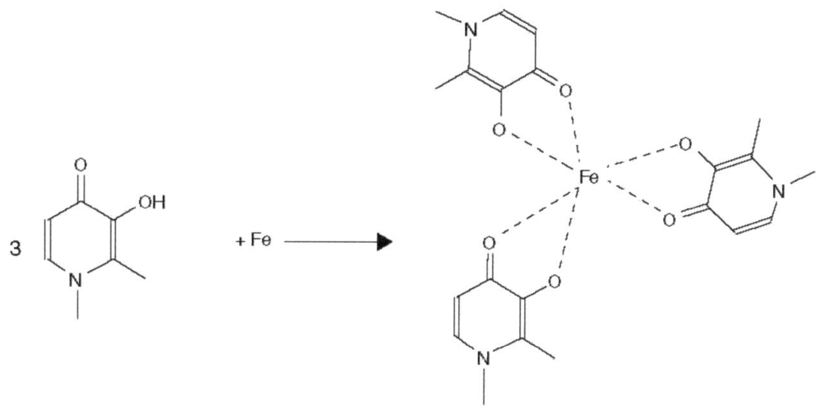

Figure 4.
Chelation of iron with Deferiprone.

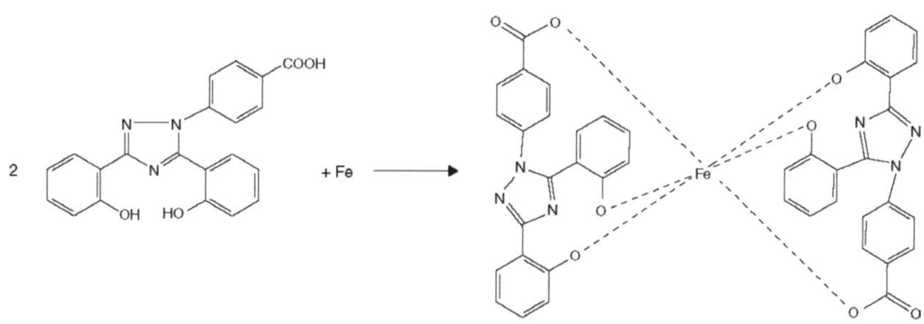

Figure 5.
Chelation of iron with Deferasirox.

The clinical DFP is used limited in thalassemia major; however, its use in agranulocytosis and arthropathy [55]. Bidentate iron chelator such as DFP requires three molecules each with two iron binding sites Fe3+ (3:1 ratio) as shown in **Figure 4**.

4.2.2.2 Deferasirox (DFS)

DFS is another synthetic oral chelator that has recently approved by US-FDA for the use in the treatment of iron overload diseases. DFS is effective in removing excess of iron from the liver. DFS has good tolerance, though monitoring is required for renal function [56]. DFS is as effective as DFO in maintaining the iron balance. The combinations of DFO and DFS or the combination of two orally active iron chelator DFP and DFS have been suggested as a treatment option for transfusional iron overload. DFS is used in the treatment of uncommon anemia like aplastic anemia, Diamond–Blackfan anemia and Fanconi's anemia, which are associated with iron overload [57]. DFS is tridentate iron chelator DFS requires two molecules for Fe3+ (2:1 ratio) as shown in **Figure 5**.

4.3 Limitations of current iron chelation therapy

All these iron chelators are having severe side effects and certain limitations. The side effects and limitations of DFO include irritation at the infusion site, growth retardation, skeletal changes, ocular and auditory disturbances, hypersensitivity reactions and systemic allergic reaction such as rash, urticaria, anaphylactic reaction, with or without shock and angioedema. DFO is an expensive drug and cannot be afforded by the majority of patients. The primary in a challenge with DFO therapy is patient's adherence. DFO is having poor oral bioavailability and short $t_{1/2}$, therefore it is administered by slow subcutaneous infusion over a period of 8–12 hrs for 5–7 days/week. This leads to lower patient compliance. The slow subcutaneous DFO infusion affect quality of life as the slow infusion can produce troublesome, time-consuming and painful. Patient's poor compliance resulted in gaps during chelation therapy, which leads to increase the plasma iron level, which causes further damage.

The major side effect of DFP is that it produces agranulocytosis, which can be reversed by discontinuation of therapy. The other side effects of DFP include gastrointestinal discomfort, arthropathy, increased liver-enzyme levels, low plasma zinc level, the progression of hepatic fibrosis associated with an increase in iron overload or hepatitis C and joint pain [58].

DFS produces various side effects such as agranulocytosis, gastrointestinal discomfort, skin rash, loss of hearing and visual impairment. Especially in geriatric patients and other patients with high risk of myelodysplastic syndrome, hepatic or renal impairment and thrombocytopenia are prone to develop hepatic failure, kidney failure and gastrointestinal hemorrhage with the use of DFS. Also, these agents are not suitable for use during pregnancy [59].

5. Herbal iron chelators

Due to these side effects and limitations, the use of synthetic iron chelators is suboptimal. Taking into account the paucity of iron chelating agents, scientists are putting their efforts towards the finding of therapeutically potential iron chelator to get maximum possible benefits with fewer harmful effects. Plants containing flavonoids and polyphenolic compounds possess iron chelating and antioxidant property [60, 61]. Due to the specific chemical structure of flavonoids, they can chelate iron and forms the soluble as well as stable iron-flavonoids complex.

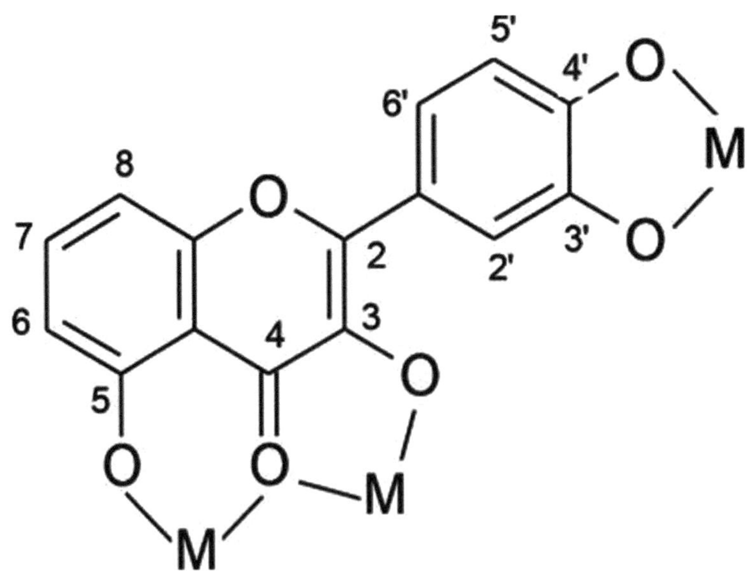

Figure 6.
Typical metal chelation sites of flavonoid.

Plant	Family	Common name	Part used	Type of study	Extract	References
Caesalpinia sappan	Caesalpiniaceae	Sappan wood	Wood	In-vivo	Hydroalcoholic	[64]
Curcuma longa	Zingiberaceae	Turmeric	rhizomes	In-Vivo	Aqueous	[65]
Triticum aestivum	Poaceae	Wheat Grass	Whole grass	In-vitro, In-vivo and clinical study	Hydroalcoholic	[66–68]
Tetracarpidium conophorum	Euphorbiaceae	African Walnut	Nut	In-vitro	Aqueous	[69]
Enhydra fluctuans	Asteraceae	Helencha	Leaves	In-vitro	Aqueous	[70]
Terminalia chebula	Combretaceae	Myrobalan	Fruit	In-vitro	Hydroalcoholic	[71]
Terminalia belerica	Combretaceae	Bahera	Fruit	In-vitro	Hydroalcoholic	[71]
Emblica officinalis	Euphorbeaceae	Indian gooseberry	Fruit	In-vitro	Hydroalcoholic	[71]
Caesalpinia crista	Caesalpiniaceae	Crested fever nut	Leaves	In-vitro	Hydroalcoholic	[71]
Cajanus cajan	Fabaceae	Pigeon pea	Leaves	In-vitro	Hydroalcoholic	[71]
Tinospora cordifolia	Menispermaceae	Heart-leaved moonseed	Stem	In-vitro	Hydroalcoholic	[71]
Medicago sativa	Fabaceae	Alfalfa	Arial	In-Vitro and In-Vivo	Aqueous and Methanol	[72–74]
Allium porrum	Alliaceae	Leek	Arial	In-Vitro and In-Vivo	Hydroalcoholic	[73, 74]
Silybum marianum	Asteraceae	Milk thistle	Seeds	In-Vivo	Aqueous and methanolic	[75, 76]
Nerium indicum	Apocynaceae	Kaner	Leaves	In-Vivo	Methanol	[77]
Clerodendrum colebrookianum	Lamiaceae	East Indian glory bowe	Leaves	In-Vivo	Aqueous and Methanol	[78]
Melilotus officinalis	Fabaceae	Yellow sweet clover	Arial	In-vitro and In-vivo	Aqueous and methanol fraction	[79]
Salvia virgata	Lamiaceae	wand sage	Shoot	In-Vivo	Dichloromethane	[76]
Epilobium hirsutum	Onagraceae	great willowherb	Leaves	In-vitro and In-vivo	Aqueous and Methanol fraction	[80]
Caulerpa racemosa	Caulerpaceae	sea grapes	Whole plant	In-Vivo	Aqueous and Ethanolic	[81]
Mangifera foetida	Anacardiaaceae	Horse mango	Leaves	In-Vivo	Aqueous	[82]
Coriandrum sativum	Apiaceae	Coriander	Whole plant	In-Vivo	Hydroalcoholic	[83]

Table 1.
Natural Iron chelating agents.

Flavonoids possess three possible sites for metal chelating that can bind metal ions as follows (**Figure 6**).

i. 3-hydroxy-4-ketone groups in the C-ring

ii. 5-hydroxy group in the A-ring and 4-carbonyl group in the C–ring

iii. 3′,4′-dihydroxy groups, located on the B-ring.

The complex is then excreted in urine and feces [62]. Flavonoids and polyphenolic compounds illustrate their antioxidant activity through various mechanisms, such as free radicals scavenging, transition metals chelation and inhibition of various enzymes [63].

Some medicinal plants were reported to have In-vitro iron chelating potential, while other medicinal plants have been screened for In-vivo iron chelation activity, whereas some plants were clinically evaluated for the treatment of iron overload in β thalassaemia patients as showed in **Table 1**.

6. Conclusion

Chelation therapy is the preferred medical treatment for reducing the toxic effects of metals. Chelating agents are capable of binding to toxic metal ions to form complex structures which are easily excreted from the body removing them from intracellular or extracellular spaces.

Presently, siderophore like DFO and synthetic iron chelators such as DFP and DFS were used for the treatment of iron overload diseases. These iron chelators have severe side effects and certain limitations. As compared to siderophores and synthetic iron chelators, natural iron chelators are usually less toxic and have minimum side effects. Additionally, these medicines possess antioxidant property, which plays an essential role in the treatment of iron overload disease and its complications associated with oxidative stress. Therefore, need to search for more safe and effective treatment of iron overloaded disease has become an area of current research interest.

Author details

Naheed Waseem A. Sheikh[1*], Satish B. Kosalge[1], Tusharbindu R. Desai[2], Anil P. Dewani[3], Deepak S. Mohale[3] and Alok S. Tripathi[4]

1 Hi-Tech College of Pharmacy, Chandrapur, Maharashtra, India

2 School of Pharmacy, RK University, Rajkot, Gujarat, India

3 P. Wadhwani College of Pharmacy, Yavatmal, Maharashtra, India

4 Amity Institute of Pharmacy, Amity University, Noida, Uttar Pradesh, India

*Address all correspondence to: wsheikh2@gmail.com

References

[1] N. Abbaspour, R. Hurrell, R. Kelishadi, Review on iron and its importance for human health, J. Res. Med. Sci. 19 (2014) 164-174.

[2] N.C. Andrews, Forging a field: the golden age of iron biology, Blood 112 (2008) 219-230, https://doi.org/10.1182/blood-2007-12-077388.

[3] M.W. Hentze, M.U. Muckenthaler, N.C. Andrews, Balancing acts: molecular control of mammalian iron metabolism, Cell 117 (2004) 285-297, https://doi.org/10.1016/s0092-8674(04)00343-5.

[4] F.M. Torti, S.V. Torti, Regulation of ferritin genes and protein, Blood 99 (2002) 3505-3516, https://doi.org/10.1182/blood.v99.10.3505.

[5] E.C. Theil, Concentrating, Storing and Detoxifying iron: the ferritin and hemosiderin, In: G. Anderson, G. McLaren, (Eds), Iron physiology and pathophysiology in humans, Nutrition and health, Humana press, Brisbane, 2012, 63-78.

[6] S.S. Nadadur, K. Srirama, A. Mudipalli, Iron transport and homeostasis mechanisms: Their role in health and disease, Indian J. Med. Res. 128 (2008) 533-544.

[7] M. Munoz, I. Villar, J.A. Garcia-Erce, An update on iron physiology, World J. Gastroentero. 15 (2009) 4617-4626, https://doi.org/10.3748/wjg.15.4617.

[8] W. Leong, B. Lonnerdal, Hepcidin, the recently identified peptide that appears to regulate iron absorption, J. Nutr. 134 (2004) 1-4, https://doi.org/10.1093/jn/134.1.1.

[9] R.H. Wang, C. Li, X. Xu, Y. Zheng, C. Xiao, P. Zerfas, S. Cooperman, M. Eckhaus, T. Rouault, L. Mishra, C. Deng, A role of SMAD4 in iron metabolism through the positive regulation of hepcidin expression, Cell Metab. 2 (2005) 399-409, https://doi.org/10.1016/j.cmet.2005.10.010.

[10] M, P, Christine and C, L, Anne, Laboratory methodologies for indicators of iron status: strengths, limitations, and analytical challenges, Am. J. Clin. Nutr. 106 (2017) 1606S-1614S https://doi.org/10.3945/ajcn.117.155887.

[11] A, Pietrangelo, Haemochromatosis, Gut 52 (2003) ii23–ii30, https://doi.org/10.1136/gut.52.suppl_2.ii23.

[12] A. Siddique, K.V. Kowdley, Review article: The iron overload syndromes, Aliment Pharmacol. Ther. 35 (2012) 876-893, https://doi.org/10.1111/j.1365-2036.2012.05051.x.

[13] G. Papanikolaou, K. Pantopoulos, Iron metabolism and toxicity, Toxicol. Appl. Pharm. 202 (2005) 199-211, https://doi.org/10.1016/j.taap.2004.06.021.

[14] D. Trinder, C. Fox, G. Vautier, J. K. Olynyk, Molecular pathogenesis of iron overload, Gut 51 (2002) 290-295, http://dx.doi.org/10.1136/gut.51.2.290.

[15] J.N. Feder, A. Gnirke, W. Thomas, Z. Tsuchihashi, D.A. Ruddy, A. Basava, F. Dormishian, R.Jr. Domingo, M.C. Ellis, A. Fullan, L.M. Hinton, N.L. Jones, B.E. Kimmel, G.S. Kronmal, P. Lauer, V.K. Lee, D.B. Loeb, F.A. Mapa, E. McClelland, N.C. Meyer, G.A. Mintier, N. Moeller, T. Moore, E. Morikang, C.E. Prass, L. Quintana, S.M. Starnes, R.C. Schatzman, K.J. Brunke, D.T. Drayna, N.J. Risch, B.R. Bacon, R.K. Wolff, A novel MHC class I-like gene is mutated in patients with hereditary haemochromatosis, Nat. Genet. 13 (1996) 399-408, https://doi.org/10.1038/ng0896-399.

[16] B.R. Bacon, L.W. Powell, P.C. Adams, T.F. Krisena, J.H. Hoofnagle, Molecular medicine and hemochromatosis: at the

crossroads, Gastroenterology 116 (1999) 193-207, https://doi.org/10.1016/S0016-5085(99)70244-1.

[17] C. Mura, O. Raguenes, C. Ferec, HFE mutations analysis in 711 hemochromatosis probands: evidence for S65C implication in mild form of hemochromatosis, Blood 93 (1999) 2502-2505, https://doi.org/10.1182/blood.V93.8.2502.

[18] J.N. Feder, Z. Tsuchihashi, A. Irrinki, V.K. Lee, F.A. Mapa, E. Morikang, C.E. Prass, S.M. Starnes, R.K. Wolff, S. Parkkila, W.S. Sly, R.C. Schatzman, The hemochromatosis founder mutation in HLA-H disrupts β2-microglobulin interaction and cell surface expression, J. Biol. Chem. 272 (1997) 14025-14028, https://doi.org/10.1074/jbc.272.22.14025.

[19] H. Zoller, R.O. Koch, I. Theurl, R.O. Koch, W. Vogel, P. Obrist, A. Pietrangelo, G. Montosi, D.J. Haile, Expression of duodenal iron transporters divalent-metal transporter 1 and ferroportin 1 in iron deficiency and iron overload, Gastroenterology 120 (2001) 1412-1419, https://doi.org/10.1053/gast.2001.24033.

[20] K.A. Stuart, G.J. Anderson, D.M. Frazer, L.W. Powell, M. McCullen, L.M. Fletcher, D.H.G. Crawford, Duodenal expression of iron transport molecules in untreated haemochromatosis subjects, Gut 52 (2003) 953-959, https://doi.org/10.1136/gut.52.7.953.

[21] A. Roetto, G. Papanikolaou, M. Politou, F. Alberti, D. Girelli, J. Christakis, D. Loukopoulos, C. Camaschella, Mutant antimicrobial peptide hepcidin is associated with severe juvenile hemochromatosis, Nat. Genet. 33 (2003) 21-22, https://doi.org/10.1038/ng1053.

[22] G. Papanikolaou, M.E. Samuels, E.H. Ludwig, M.L.E. MacDonald, P.L. Franchini, M. Dube, L. Andres, J. MacFarlane, N. Sakellaropoulos, M. Politou, E. Nemeth, J. Thompson, J.K. Risler, C. Zaborowska, R. Babakaiff, C.C. Radomski, T.D. Pape, O. Davidas, J. Christakis, P. Brissot, G. Lockitch, T. Ganz, M.R. Hayden, Y.P. Goldberg, Mutations in HFE2 cause iron overload in chromosome 1q-linked juvenile hemochromatosis, Nat. Genet. 36 (2004) 77-82, https://doi.org/10.1038/ng1274.

[23] C. Camaschella, A. Roetto, A. Cali, M.D. Gobbi, G. Garozzo, M. Carella, N. Majorano, A. Totaro, P. Gasparini, The gene TFR2 is mutated in a new type of haemochromatosis mapping to 7q22, Nat. Genet. 25 (2000) 14-15, https://doi.org/10.1038/75534.

[24] G. Montosi, A. Donovan, A. Totaro, C. Garuti, E. Pignatti, S. Cassanelli, C.C. Trenor, P. Gasparini, N.C. Andrews, A. Pietrangelo, Autosomal-dominant hemochromatosis is associated with a mutation in the ferroportin (SLC11A3) gene, J. Clin. Invest. 108 (2001) 619-623, https://doi.org/10.1172/JCI13468.

[25] A. Hayashi, Y. Wada, T. Suzuki, A. Shimizu, Studies on familial hypotransferrinemia: unique clinical course and molecular pathology, Am. J. Hum. Genet. 53 (1993) 201-213.

[26] P. Pootrakul, K. Kitcharoen, P. Yansukon, P. Wasi, S. Fucharoen, P. Charoenlarp, G. Brittenham, M.J. Pippard, C.A. Finch, The effect of erythroid hyperplasia on iron balance, Blood 71 (1988) 1124-1129, https://doi.org/10.1182/blood.V71.4.1124.1124.

[27] A.V. Hoffbrand, F. Al-Refaie, B. Davis, N. Siritanakatkul, B.F.A. Jackson, J. Cochrane, E. Prescott, B. Wonke, Long term trial of deferiprone in 51 transfusion-dependent iron overloaded patients, Blood 91 (1998) 295-300, https://doi.org/10.1182/blood.V91.1.295.

[28] V. Gordeuk, J. Mukiibi, S.J. Hasstedt, W. Samowitz, C.Q. Edwards, G. West, S. Ndambire, J. Emmanual, N. Nkanza, Z. Chapanduka, M. Randall, P. Boone, P. Romano, R.W. Martell, T. Yamashita, P.

Effler, G. Brittenham, Iron overload in Africa: Interaction between a gene and dietary iron content, N. Engl. J. Med. 326 (1992) 95-100, https://doi.org/10.1056/NEJM199201093260204.

[29] V.R. Gordeuk, A. Caleffi, E. Corradini, F. Ferrara, R.A. Jones, O. Castro, O. Onyekwere, R.Kittles, E. Pignatti, G. Montosi, C. Garuti, I.T. Gangaidzo, Z.A.R. Gomo, V.M. Moyo, T.A. Rouault, P. MacPhail, A. Pietrangelo, Iron overload in Africans and African-Americans and a common mutation in SCL40A1 (ferroportin 1) gene, Blood Cells Mol. Dis. 31 (2003) 299-304, https://doi.org/10.1016/S1079-9796(03)00164-5.

[30] J. Ludwig, E. Hashimoto, M.K. Porayko, T.P. Moyer, W.P. Baldus, Hemosiderosis in cirrhosis: a study of 447 native livers, Gastroenterology 112 (1997) 882-888, https://doi.org/10.1053/gast.1997.v112.pm9041250.

[31] B.Y. Tung, M.J. Emond, M.P. Bronner, S.D. Raaka, K.V. Kowdley, Hepatitis C, iron status and disease severity: relationship with HFE mutations, Gastroenterology 124 (2003) 318-326, https://doi.org/10.1053/gast.2003.50046.

[32] R.G. Batey, R. Johnston, Effects of alcohol, carbon tetrachloride and choline deficiency on iron metabolism in the rat, Alcohol. Clin. Exp. Res. 17 (1993) 931-934, https://doi.org/10.1111/j.1530-0277.1993.tb05643.x.

[33] A. Pascoe, P. Kerlin, C. Steadman, A. Clouston, D. Jones, L. Powell, E. Jazwinska, S. Lynch, R. Strong, Spur cell anaemia and hepatic iron stores in patients with alcoholic liver disease undergoing orthotopic liver transplantation, Gut 45 (1999) 301-305, https://doi.org/10.1136/gut.45.2.301.

[34] C.K. Ong, S.L. Lim, W.C. Tan, E.E. Ong, A.S. Goh, Endocrine complications in transfusion dependent thalassaemia in penang hospital, Med. J. Malaysia 63 (2008) 109-112.

[35] E. Dimitriou, M. Kairis, J. Sarafidou, H. Michelakakis, Iron overload and kidney lysosomes, Biochim. Biophys. Acta 1501 (2000) 138-148, https://doi.org/10.1016/s0925-4439(00)00019-3.

[36] H.M. Schipper, Neurodegeneration with brain iron accumulation - Clinical syndromes and neuroimaging, Biochim. Biophys. Acta 1822 (2012) 350-360, https://doi.org/10.1016/j.bbadis.2011.06.016.

[37] D.S. Kalinowski, D.R. Richardson, The evolution of iron chelators for the treatment of iron overload disease and cancer, Pharmacol. Rev. 57 (2005) 547-583, https://doi.org/10.1124/pr.57.4.2.

[38] Z.D. Liu, R.C. Hider, Design of clinically useful iron (III)-selective chelators, Med. Res. Rev. 22 (2002) 26-64, https://doi.org/10.1002/med.1027.

[39] R.C. Hider, T. Zhou, The design of orally active iron chelators, Ann. NY Acad. Sci. 1054 (2005) 141-154, https://doi.org/10.1196/annals.1345.017.

[40] J.L. Buss, F.M. Torti, S.V. Torti, The role of iron chelation in cancer therapy, Curr. Med. Chem. 10 (2003) 1021-1034, https://doi.org/10.2174/0929867033457638.

[41] J.B. Porter, Evaluation of new iron chelators for clinical use, Acta Haematol. 95 (1996) 13-25, https://doi.org/10.1159/000203852.

[42] N.T. Le, D.R. Richardson, Competing pathways of iron chelation: angiogenesis or anti-tumor activity: targeting different molecules to induce specific effects, Int. J. Cancer 110 (2004) 468-469, https://doi.org/10.4155/fmc.09.121.

[43] Y. Deugnier, P. Brissot, O. Loreal, Iron and the liver: update 2008, J. Hepatol. 48 (2008) S113-S123, https://doi.org/10.1016/j.jhep.2008.01.014.

[44] T.Y. Tang, S.P. Howarth, S.R. Miller, M.J. Graves, J.M. U-King-Im, Z.Y. Li, S.R.

Walsh, P.D. Hayes, K. Varty, J.H. Gillard, Comparison of the inflammatory burden of truly asymptomatic carotid atheroma with atherosclerotic plaques in patients with asymptomatic carotid stenosis undergoing coronary artery bypass grafting: an ultrasmall super-paramagnetic iron oxide enhanced magnetic resonance study, Eur. J. Vasc. Endovasc. Surg. 35 (2008) 392-398, https://doi.org/10.1016/j.ejvs.2007.10.019.

[45] J.L. Sullivan, Iron in arterial plaque: a modifiable risk factor for atherosclerosis, Biochim. Biophys. Acta 1790 (2009) 718-723, https://doi.org/10.1016/j.bbagen.2008.06.005.

[46] S. Swaminathan, V.A. Fonseca, M.G. Alam, S.V. Shah, The role of iron in diabetes and its complications, Diabetes Care 30 (2007) 1926-1933, https://doi.org/10.2337/dc06-2625.

[47] L. Li, B. Frei, Prolonged exposure to LPS increases iron, heme, and p22phox levels and NADPH oxidase activity in human aortic endothelial cells: inhibition by desferrioxamine, Arterioscler. Thromb. Vasc. Biol. 29 (2009) 732-738, https://doi.org/10.1161/ATVBAHA.108.183210.

[48] A.S. Ibrahim, B. Spellberg, J.Jr. Edwards, Iron acquisition: a novel perspective on mucormycosis pathogenesis and treatment, Curr. Opin. Infect. Dis. 21 (2008) 620-625, https://doi.org/10.1097/QCO.0b013e3283165fd1.

[49] R.C. Hider, Y. Ma, F. Molina-Holgado, A. Gaeta, S. Roy, Iron chelation as a potential therapy for neurodegenerative disease. Biochem. Soc. Trans. 36 (2008) 1304-1308, https://doi.org/10.1042/BST0361304.

[50] H. Haas, M. Eisendle, B.G. Turgeon, Siderophores in fungal physiology and virulence, Annu. Rev. Phytopathol. 46 (2008) 149-187, https://doi.org/10.1146/annurev.phyto.45.062806.094338.

[51] C. Vermylen, What is new in iron overload?, Eur. J. Pediatr. 167 (2008) 377-381, https://doi.org/10.1007/s00431-007-0604-y.

[52] J.C. Barton, Optimal management strategies for chronic iron overload, Drugs 67 (2007) 685-700, https://doi.org/10.2165/00003495-200767050-00004.

[53] B.J. Reeder, R.C. Hider, M.T. Wilson, Iron chelators can protect against oxidative stress through ferryl heme reduction, Free Radic. Biol. Med. 44 (2008) 264-273, https://doi.org/10.1016/j.freeradbiomed.2007.08.006.

[54] G.J. Kontoghiorghes, K. Pattichi, M. Hadjigavriel, A. Kolnagou, Transfusional iron overload and chelation therapy with deferoxamine and deferiprone (L1), Transfus. Sci. 23 (2000) 211-223, https://doi.org/10.1016/S0955-3886(00)00089-8.

[55] T.E. Delea, J. Edelsberg, O. Sofrygin, S.K. Thomas, J.F. Badali, P.D. Phatak, T.D. Coates, Consequences and costs of noncompliance with iron chelation therapy in patients with transfusion-dependent thalassemia: a literature review. Transfusion 47 (2007) 1919-1929, https://doi.org/10.1111/j.1537-2995.2007.01416.x.

[56] M.D. Cappellini, A. Cohen, A. Piga, M. Bejaoui, S. Perrotta, L. Agaoglu, Y. Aydinok, A. Kattamis, Y. Kilinc, J. Porter, M. Capra, R. Galanello, S. Fattoum, G. Drelichman, C. Magnano, M. Verissimo, M. Athanassiou-Metaxa, P. Giardina, A. Kourakli-Symeonidis, G. Janka-Schaub, T. Coates, C. Vermylen, N. Olivieri, I. Thuret, H. Opitz, C. Ressayre-Djaffer, P. Marks, D. Alberti, A Phase 3 study of deferasirox (ICL670), a once-daily oral iron chelator, in patients with β-thalassemia, Blood 107 (2006) 3455-3462, https://doi.org/10.1182/blood-2005-08-3430.

[57] H. Nick, Iron chelation, quo vadis?, Curr. Opin. Chem. Biol. 11 (2007)

419-423, https://doi.org/10.1016/j.cbpa.2007.04.025.

[58] A.N. Saliba, A.R. Harb, A.T. Taher, Iron chelation therapy in transfusion-dependent thalassemia patients: current strategies and future directions, J. Blood Med. 6 (2015) 197-209, https://doi.org/10.2147/JBM.S72463.

[59] J.B. Porter, Practical management of iron overload, Brit. J. Haematol. 115 (2001) 239-252, https://doi.org/10.1046/j.1365-2141.2001.03195.x.

[60] M. Symonowicz, M. Kolanek, Flavonoids and their properties to form chelate complexes, Biotechnol. Food Sci. 76 (2012) 35-41.

[61] M.A. Ebrahimzadeh, F. Pourmorad, A.R. Bekhradnia, Iron chelating activity, phenol and flavonoid content of some medicinal plants from Iran, Afr. J. Biotechnol. 7 (2008) 3188-3192, https://doi.org/10.5897/AJB08.476.

[62] C. Hershko, Determinants of fecal and urinary iron excretion in desferrioxamine treated rats, Blood 51 (1978) 415-423, https://doi.org/10.1182/blood.V51.3.415.415.

[63] P. Cos, T.D. Bruyne, N. Hermans, S. Apers, D.V. Berghe, A.J. Vlietinck, Proanthocyanidins in health care: current and new trends, Curr. Med. Chem. 11 (2004) 1345-1359, https://doi.org/10.2174/0929867043365288.

[64] R. Safitri, A.M. Maskoen, M.R.A.A. Syamsunarno, M. Ghozali, R. Panigoro, Iron chelating activity of Caesalpinia sappan L. extract on iron status in iron overload rats (*Rattus norvegicus* L.), AIP Conf. Proc. 2002, 020050-1-020050-6, https://doi.org/10.1063/1.5050146.

[65] Y. Jiao, J. Wilkinson IV, E.C. Pietsch, J.L. Buss, W. Wang, R. Planalp, F.M. Torti, S.V. Torti, Iron chelation in the biological activity of curcumin, Free Radic. Biol. Med. 40 (2006) 1152-1160, https://doi.org/10.1016/j.freeradbiomed.2005.11.003.

[66] T.R. Desai, J.K. Solanki, P. Buch, R.K. Goyal, Triticum aestivum (wheatgrass) formulation: An alternate treatment for the patients with Thalassemia, Ori. Pharm. Exp. Med. 7 (2008) 466-476, https://doi.org/10.3742/OPEM.2008.7.5.466.

[67] P.R. Tirgar, T.R. Desai, Investigation into iron chelating activity of Triticum aestivum (wheat grass) in iron-dextran induce iron overload model of thalassaemia, J. Pharm. Res. 4 (2011) 3066-3069.

[68] P. Das, A. Mukhopadhyay, S. Mandal, B.C. Pal, R. Mishra, D. Mukherjee, S. Mukhopadhyay, J. Basak, M. Kar, In vitro studies of iron chelation activity of purified active ingredients extracted from Triticum aestivum Linn. (Wheat Grass), Eur. J. Med. Plants 2 (2012) 113-124.

[69] B.M. Olabinri, O.O. Eniyansoro, C.O. Okoronkwo, P.F. Olabinri, M.T. Olaleye, Evaluation of chelating ability of aqueous extract of Tetracarpidium conophorum (African walnut) in vitro, Int. J. Appl. Res. Nat. Prod. 3 (2010) 13-18.

[70] L.N. Patralekh, G. Mukherjee, In vitro studies on antioxidant and iron-chelating activity of Enhydra fluctuans Lour, Sci. Cult. 76 (2010) 537-539.

[71] R. Sarkar, N. Mandal, Study of iron chelating and DNA protective activities in hydroalcoholic extract of Indian medicinal plants, Int. J. Pharm. Bio. Sci. 3 (2012) 282-291.

[72] R. Patel, P. Tirgar, Evaluation of beneficial effects of Medicago sativa (Alfalfa) in iron-overload conditions, J. Chem. Bio. Phy. Sci. Sec. B. 3 (2013) 2628-2643.

[73] A. Mirzaei, M. Abbasi, S. Sepehri, M. Mirzaei, The effects of Allium

porrum and Medicago sativa on iron concentration in thalassemia serums, Life Sci. J. 10 (2013) 27-31, https://doi.org/10.7537/marslsj1011s13.05.

[74] A. Mirzaei, H. Delaviz, M. Mirzaei, M. Tolooei, The effects of Medicago Sativa and Allium Porrum on iron overload in Rats, Global J. Health Sci. 7 (2015) 137-142, https://doi.org/10.5539/gjhs.v7n7p137.

[75] M. Nematbakhsh, Z. Pezeshki, B.A. Moaeidi, F. Eshraghi-Jazi, A. Talebi, H. Nasri, S. Baradaran, M. Gharagozloo, T. Safari, M. Haghighi, Protective role of Silymarin and Deferoxamine against iron dextran-induced renal iron deposition in male rats, Int. J. Prev. Med. 4 (2013) 286-292.

[76] M.A. Ebrahimzadeh, M. Khalili, M. Azadbakht, M. Azadbakh, Salvia virgata Jacq. and Silibum marianum L. gaertn display significant iron-chelating activity, Inter. J. Pharm. Sci. Res. 7 (2016) 3756-3763, https://doi.org/10.13040/IJPSR.0975-8232.7(9).3756-63.

[77] N.B. Ghate, D. Chaudhuri, S. Panja, N. Mandal, Nerium indicum leaf alleviates iron-induced oxidative stress and hepatic injury in mice, Pharm. Biol. 53 (2015) 1066-1074, https://doi.org/10.3109/13880209.2014.959612.

[78] A. Das, D. Chaudhuri, N.B. Ghate, S. Panja, A. Chatterjee, N. Mandal, Protective effect of Clerodendrum colebrookianum leaves against iron-induced oxidative stress and hepatotoxicity in Swiss albino mice, Indian J. Exp. Biol. 53 (2015) 281-291.

[79] N.A. Sheikh, T.R. Desai, P.R. Tirgar, Investigation into iron chelating and antioxidant potential of Melilotus officinalis in iron dextran induced iron overloaded sprague dawley rat model, Drug Res. 66 (2016) 618-627, https://doi.org/10.1055/s-0042-113182.

[80] N.A. Sheikh, T.R. Desai, P.R. Tirgar, Evaluation of iron chelating and antioxidant potential of Epilobium hirsutum for the management of iron overload disease, Biomed. Pharmacother. 89 (2017) 1353-1361, https://doi.org/10.1016/j.biopha.2017.02.079.

[81] E.K.A. Adham, A.I. Hassan, S.M. Hamed, A.A. Saber, Evaluation of iron-chelating activity of Caulerpa racemosa in iron-dextran induced iron overload in an experimental model of thalassemia, Int. J. Clin. Exp. Med. 10 (2017) 5561-5577.

[82] P. Fajri, A. Estuningtyas, M. Louisa, H. Freisleben, The preventive effect of Mangifera foetida L. leaf extract administered simultaneously to excess iron on markers of iron overload in sprague dawley rats, Med. J. Indones. 26 (2017) 246-252, http://dx.doi.org/10.13181/mji.v26i4.1829.

[83] R. Talaei, A. Kheirollah, H.B. Rezaei, E. Mansouri, G. Mohammadzadeh, Protective effects of hydro-alcoholic extract of Coriandrum sativum in rats with experimental iron-overload condition, Jundishapur J. Nat. Pharm. Prod. 13 (2018) 1-9, https://doi.org/10.5812/jjnpp.65028.

Chapter 5

Chromium Genotoxicity Associated with Respiratory Disease

Jyoti Kant Choudhari, Jyotsna Choubey, Mukesh Kumar Verma, Anand Kumar Jayapal and Biju Prava Sahariah

Abstract

Chromium existing in the biosphere in prominent two forms Cr (III) and Cr (VI) is a well-studied heavy metal. Cr (III) is considered as non-harmful and necessary element in diet whereas Cr(VI) is extremely toxic exerting various negative health impacts on human and other organisms. Mining activity is must for extracting economic minerals and a large number of people are related to these sites as worker or habitants and a major source of chromium exposure. Present chapter discusses genotoxic nature of chromium considering respiratory disease resulted from chromium exposure. The genotoxicity is illustrated in terms of chromium induced differential expressed genes (DEGs), transcription factors and microRNA regulating the DEGs and their gene ontology.

Keywords: Mine tailing, Chromium Toxicity, Genotoxicity, Gene expressions

1. Introduction

For the growth of economy of a country and improving living status of population, industrial functioning is mandatory which is in other hand associated activities including supply of power, raw materials, processing and discharge of waste. For a major section of industries, power supply is from coal or electricity generated from coal, and the raw materials are various form of ores received from mining. Mine tailing is the fine residual mine dump after completion of mining left with dug out soil, scattered residuals and disturbed ecosystem. The major source of chromium in the mine tailings is the residual ores present in traces not extracted with economic point of view and mineral processing chemicals that are left unattended. Chromium (Cr), a valuable element often finds its utility in metallurgical, chemical, and refractory industries due to its pigment property, hardness and persistence. From environment point of view, chromium exists in three oxidative states, elemental chromium (0) that does not exist naturally, whereas trivalent chromium (Cr III) is rather stable followed by hexavalent chromium (Cr VI) based on the different number of electrons and therefore varied properties [1]. Hexavalent chromium is extremely toxic even in low concentration and listed as carcinogenic, hematotoxic and altering genetic material whereas, Cr (III) is regarded as micronutrient in human diet. When Cr is left unattended in mine tailings, it can be transported by natural means to nearby waterbody, added with acid mine drainage, and surrounding ecosystem expanding the circumference of toxicity exposure [2]. This chapter emphasises on toxicity of hexavalent chromium in genetic level that influence expression of genes, the transcript factors controlling the differentially expressed genes and finally to find out the major indicating and influenced genetic factors with functional analysis of gene ontology for respiratory units of human.

1.1 Source and toxicity of chromium

Toxicity of chromium is directly influenced by the chromium species with valence with number of electrons and thus their properties. The Cr(VI), is a powerful oxidizing agent and plainly toxic to human and other organisms causing adverse effect to blood cells, renal cells, allergic conditions and organs of most part of body failure. Chromium can significantly find its route of exposures through dermal chromium contact in waste sites, inhalation of chromium emissions and ingestion of contaminated water or food grown in chromium contaminated soil. Also, erosion products and emissions from road and cement dust, leather, paints and or any Cr used materials contribute to inhalation of chromium. Dermal ulcers, irritation and sensitization of respiratory/lungs are consecutive result of chromium contact. In the plasma and cells, Cr(VI) readily get reduced to Cr(III), and thereafter excreted in the urine. Trivalent chromium is the form of chromium that is essential to human health and counted as an essential trace mineral in the human diet. Hexavalent chromium is recognised as genotoxic as it can damage genetic information in living cells, causes DNA mutations, and possibly the formation of cancerous tumours. Chromates (chromium salts) formed from hexavalent chromium also finds utilization in manufacture leather products, paints, cement, mortar, anti-corrosives, and other things. They are carcinogenic and allergenic.

1.2 Physiologic effects of chromium exposure in respiratory disease

Occupational exposures often include mixed exposure to both Cr(III) and Cr (VI) [3]. Chromium compounds, when inhaled, causes respiratory tract irritants, resulting in airway irritation, airway obstruction, and lung, nasal, or sinus cancer. Radiographic analysis from several reports revealed enlargement of the hilar region and lymph nodes [4, 5]. Consistent associations have been found between employment in the chromium industries and significant risk for respiratory cancer. Moller et al. [6] reported systemic reactions characterised with anaphylactoid reaction in a young welder having chromium (VI) vapor fume exposures. Following an experiment with sodium chromate inhalation at a concentration of 29 $\mu g/m^3$, formation of static urticaria, angioedema and severe bronchospasm simultaneously with plasma histamine rising in threefold was documented and suggested direct positive leukocyte inhibitory factor of sodium chromate.

A number of nasal mucosa injury cases in Cr (VI) exposed workers at concertation of nearly 20 $\mu g/m^3$ (against US permissible standard 5 $\mu g/m^3$) for 5 months to 10 years characterised with inflamed mucosa and ulcerated/perforated septum was recorded in a study with 43 chrome-plating plants and tanneries in Sweden [7, 8]. Huge number of complaints for nasal irritations was documented in a detail epidemiological study with Tokyo (Japan) housewives residing near chromium slag contaminated construction site [7]. U.S has recommended chromate and chromic acid at workplace to be 5 $\mu g/m^3$ as permissible standard. Gibb et al. [9] observed that with less than 30 days median time for nasal ulceration diagnosis from first exposure, median Cr (VI) concentration matched the Sweden report. Occupational exposure to Cr(III) has also been associated with respiratory effects.

Persons developed coughing, wheezing, and decreased forced volume after an inhalation exposure to a sample of Cr(III) sulfate [10]. Combine effect of Cr(III) and Cr(VI) as total chromium (0.02–0.19 mg total chromium/m^3) investigated among 60 ferrochromium workers squeezed out subjective symptoms of coughing, wheezing, and dyspnea whereas control remained neutral [11]. These symptoms might get puzzled with smoking issue to clarify the accurate problem of the diseases [11]. While considering respiratory issue, animals are also often exposed to chromium similar to the human. Henderson et al. [12] in histological examination with exposure of 0.9–25 mg Cr(III) trichloride for 30 min observed alterations in lung tissues associated with mild inflammation.

2. Chromium-gene interactions in respiratory disease

Comparative toxico-genomics database (CTD, http://ctdbase.org) is a recognised well informed/updated, openly accessible database. It purposes to provide detail knowledge and information about the impacts of exposure of environmental elements (pollutants) on human health.

The core block of the database basically manually curated contains updated information regarding interaction and relationships among chemicals, genes, proteins and their resulted specific disease in terms of functional and pathways to incorporate new hypotheses expressing underlying mechanisms of disease and environmental contamination [13].

3. Results and discussion

In this work, all Chromium- gene /protein interactions for respiratory disease are downloaded from CTD, in which Chromium- gene /protein interactions associated to the following 04 respiratory disease are selected for further analysis according to MESH ID used in CTD— Lung Neoplasma, Pulmonary Fibrosis and Lung disease. Chromium- gene/protein interactions associated to this respiratory disease are collected for further analysis. According to the reference score on relationships between chemicals-genes, genes-diseases and chemicals-diseases [14], lung neoplasms is recognised as most likely having the maximum connectivity with chromium. (**Table 1**). From the identified 168 chromium gene with in respiratory disease, 131 genes are unique.

3.1 Gene function enrichment analysis

KEGG (http://www.genome.jp/) is a knowledge base for systematic analysis of gene functions, linking genomic information with higher-order functional information [15]. For the analysis of Gene Ontology (GO) and Kyoto Encyclopaedia of Genes and Genomes (KEGG) pathway analysis, the Database for Annotation, Visualization and Integrated Discovery (DAVID, http://david.abcc.ncifcrf.gov/) is a great option. DAVID provides various functional annotation tools for researchers to understand biological meaning behind large list of genes. [16] Gene ontology (GO) analysis and Kyoto Encyclopaedia of Genes and Genomes (KEGG) pathway enrichment analysis can be performed for analysing differentially expressed genes (DEGs) at the functional level based on DAVID Bioinformatics Resources 6.8. $P < 0.05$ as the cut-off criterion. Researchers can upload all DEGs to the online software DAVID to identify overrepresented GO categories and KEGG pathways. The curated genes in CTD for each respiratory disease can be uploaded to DAVID 6.8 Beta

Disease Name	Inference chromium-interacted genes (n)	Gene count	Inference Score
Lung Neoplasms (Cr VI)	ACE,AKT1,APOA1,APOC3,AR,AVPI1,BCL2L1,BRCA2,CASP8, CCND1,CDKN1A,CDKN1B,CDKN2A,CEACAM1,CHEK2, COL6A1,COX17,CRP,CTNNB1,CWH43,CYP1B1,DPYD,EEF2, EFNB2,EGFR,EGR1,ERBB2,ERBB3,ESR1,FAS,FEN1,FGF9,FOS, GCLC,GPX1,GPX3,GSTM1,GSTP1,HILPDA,HMOX1,HRAS, IDS,IER2,IFNG,IL1B,IL2,IL6,JUN,JUNB,LECT2,MAP2K7, MAPK1,MAPK14,MAPK3,MIR21,MIR494,MMP10,MYC,NOS2, OGG1,PCNA,PDCD4,PRDX1,PRDX6,PRKN,PTMA,SERPINA1, SERPING1,SFTPB,SIDT2,SMC2,SOX2,SOX9,TERT,TFRC, TGFBR2,TNF,TP53,TRP53,USP18,WNT5A	81	74.7
Lung Neoplasms(Cr III)	ACE,AKT1,ANXA2,APOA1,AZGP1,CASP8,CAV1,CDKN1A, CDKN2A,CPE,CYP1B1,FOS,GCLC,GJA1,GPX1,GSTM1, HMOX1,IER2,IFNG,IL10,IL1B,IL6,JUN,MAPK1,MAPK3,MMP1, SFTPB,TGFB1,TLR4,TNF,TP53,TYMS	32	58.22
Pulmonary Fibrosis	ACE2, ACTA2, CAT, CCL11, CCL2, CCL5, CXCL8, EDN1, FAM13A, FN1, FYN, HMGB1, HMOX1, IL1B, IL4, IL6, LAMB1, MMP2, MMP9, MTOR, NFE2L2, PARP1, PDGFB, PTX3, SERPINA1, SOD1,STAT3,TIMP1,TNF	29	25.25
Lung Diseases	ACE, BST1, HARS, HIF1A, IGF1R, INSR, KIT, PDGFRA,PTGS2, SERPINA1,SFTPB,SOD2,TNF,VEGFA	14	7.62
Asthma, Occupational	TGFB1, TNF	02	5.86
Lung Injury	ACE, ACE2, CCL2, CYP1A1, HMOX1, IL6, PARP1, SIRT1, TNF	09	3.12
Nose Neoplasms	MMP2	01	2.55

Table 1.
Selected Respiratory diseases and related chromium-interacted gene.

(https://david-d.ncif-crf.gov/tools.jsp) with *Homo sapiens* as the background population [17] for GO analysis.

GO analysis results for Cr toxicity in respiratory organs shows that that chromium interacted genes in respiratory disease are involved in the biological processes (BP) such as positive regulation of gene expression, positive regulation of cell proliferation, response to drug positive regulation of protein phosphorylation. (**Table 2**) For molecular function (MF), genes are enriched in identical protein binding, enzyme binding, transcription factor binding and protein phosphatase binding (**Table 2**). In addition, GO cell component (CC) analysis also displayed that the gene are significantly enriched in the extracellular space, protein complex, extracellular region and extracellular exosome (**Table 2**).

Table 3 contains the most significantly enriched pathways of the chromium interacting genes by KEGG analysis. The interacting genes are enriched in Pathways in cancer, Proteoglycans in cancer, HIF-1 signalling pathway and TNF signalling pathways.

3.2 Gene-TFs-miRNAs regulation

The transcription factors (TFs) as well as microRNAs (miRNAs), are recognised for their huge share in transacting and gene regulations with various common logics and regulatory factors for gene regulation in multicellular genomes [18, 19]. The library of ENCODE and ChEA Consensus TFs from ChIP-X in EnrichR (http://amp.pharm.mssm.edu/Enrichr/ [20, 21]) can be used for the possible TFs

Term	Count	P value	FDR	Genes
Biological Process				
Positive regulation of gene expression	26	1.17E-21	2.49E-18	CRP, PDGFB, HIF1A, TNF, GJA1, FGF9, ERBB3, MYC, ERBB2, HRAS, TGFB1, CAV1, STAT3, FN1, MAPK14, MTOR, VEGFA, ACTA2, AR, IL6, IFNG, IL1B, KIT, TP53, TLR4, NFE2L2
Positive regulation of nitric oxide biosynthetic process	14	1.39E-18	1.49E-15	EDN1, INSR, PTGS2, SOD2, ESR1, TNF, EGFR, MTOR, IL6, IFNG, IL1B, AKT1, PTX3, TLR4
Positive regulation of cell proliferation	26	1.23E-15	8.80E-13	CDKN1B, HILPDA, PDGFB, EGFR, IGF1R, EFNB2, FGF9, MYC, MAPK1, SOX9, TIMP1, HRAS, PDGFRA, EDN1, TGFB1, INSR, STAT3, FN1, IL2, TGFBR2, VEGFA, AR, IL6, IFNG, KIT, BCL2L1
Aging	18	1.68E-15	8.89E-13	JUN, TGFB1, OGG1, STAT3, FOS, TYMS, EEF2, TGFBR2, SOD1, GCLC, IL6, CAT, CYP1A1, SERPING1, CCL2, AKT1, TIMP1, NFE2L2
Response to drug	22	2.08E-15	8.89E-13	CDKN1A, JUN, TGFB1, CDKN1B, OGG1, STAT3, APOA1, FOS, TYMS, PTGS2, SOD2, TGFBR2, SOD1, IL4, IL6, IFNG, CCND1, MYC, CAT, CYP1A1, CTNNB1, FYN
Positive regulation of smooth muscle cell proliferation	13	7.81E-15	2.78E-12	JUN, EDN1, PDGFB, PTGS2, TNF, EGFR, MTOR, TGFBR2, IL6, MYC, CCL5, AKT1, HMOX1
Positive regulation of protein phosphorylation	16	1.07E-14	3.27E-12	TGFB1, ANXA2, INSR, TNF, MMP9, EGFR, MTOR, VEGFA, CCND1, CHEK2, IL1B, ERBB2, AKT1, SOX9, HRAS, MAPK3
Term	Count	P value	FDR	Genes
Cellular Component				
Extracellular space	49	1.61E-24	3.37E-22	SERPINA1, CXCL8, TFRC, HILPDA, LECT2, HMGB1, TNF, FGF9, TIMP1, SFTPB, EDN1, ANXA2, GPX3, MMP2, WNT5A, APOA1, MMP9, MMP10, ACTA2, ACE2, AZGP1, IFNG, IL1B, CAT, KIT, SERPING1, CRP, CCL11, GSTP1, PDGFB, EGFR, ERBB3, PRDX1, CCL5, CCL2, HMOX1, TGFB1, ACE, FN1, APOC3, LAMB1, PRDX6, IL2, SOD1, VEGFA, IL4, IL6, CPE, PTX3
Protein complex	21	1.48E-12	1.55E-10	PDGFRA, CDKN1A, FEN1, CDKN1B, PARP1, CDKN2A, OGG1, CAV1, BRCA2, PTGS2, SOD1, ACTA2, AR, MYC, COL6A1, AKT1, MAPK1, CTNNB1, SOX9, TP53, MAPK3
Extracellular region	37	9.09E-12	6.36E-10	CRP, SERPINA1, CCL11, CXCL8, TFRC, PDGFB, HMGB1, TNF, FGF9, CCL5, CCL2, TIMP1, SFTPB, EDN1, TGFB1, ACE, GPX3, MMP2, WNT5A, FN1, APOA1, APOC3, LAMB1, MMP9, MMP10, IL2, SOD1, VEGFA, IL4, ACE2, IL6, AZGP1, IFNG, IL1B, COL6A1, SERPING1, PTX3
Cytosol	49	4.84E-09	2.54E-07	CDKN1A, CDKN1B, FAM13A, SMC2, SOX2, GJA1, CASP8, CCND1, MYC, AKT1, HRAS, GPX1, ANXA2, APOA1, FOS, TGFBR2,

Term	Count	P value	FDR	Genes
				ACTA2, AR, IL1B, DPYD, CAT, TP53, GSTP1, TYMS, USP18, HIF1A, PRDX1, HMOX1, MAPK1, FYN, HARS, MAP2K7, MAPK3, JUN, GSTM1, NOS2, CDKN2A, STAT3, EEF2, MAPK14, PRDX6, MTOR, SOD1, GCLC, PDCD4, CTNNB1, FAS, NFE2L2, BCL2L1
Mitochondrion	26	7.77E-07	2.62E-05	FEN1, GSTP1, OGG1, COX17, TYMS, GJA1, CASP8, MYC, PRDX1, CYP1B1, AKT1, MAPK1, FYN, HARS, MAPK3, GPX1, PARP1, CDKN2A, MMP2, MAPK14, SOD2, SOD1, CAT, CYP1A1, TP53, BCL2L1
Membrane raft	11	7.89E-07	2.62E-05	ACE2, GJA1, CASP8, ANXA2, CAV1, FAS, FYN, EEF2, TNF, EGFR, TGFBR2
Extracellular exosome	40	8.72E-07	2.62E-05	CRP, SERPINA1, PCNA, TFRC, GSTP1, SMC2, GJA1, FGF9, PRDX1, MAPK1, TIMP1, MAPK3, ACE, GPX1, ANXA2, GPX3, INSR, WNT5A, FN1, APOA1, APOC3, LAMB1, EEF2, MAPK14, SOD2, MMP9, PRDX6, SOD1, ACTA2, ACE2, BST1, AZGP1, CEACAM1, IL1B, COL6A1, CAT, SERPING1, CPE, CTNNB1, FAS
Molecular Function				
Identical protein binding	31	2.03E-15	8.34E-13	SERPINA1, PCNA, TFRC, LECT2, PDGFB, TNF, EGFR, IGF1R, CASP8, ERBB3, CHEK2, PRDX1, ERBB2, AKT1, MAPK1, FYN, JUN, PARP1, CAV1, STAT3, FN1, APOA1, SOD2, MMP9, ESR1, SOD1, VEGFA, FAS, PTX3, TP53, BCL2L1
Enzyme binding	22	9.84E-15	2.02E-12	JUN, TGFB1, GSTM1, PCNA, PARP1, CAV1, APOA1, PTGS2, MAPK14, HIF1A, ESR1, EGFR, AR, CCND1, CAT, CYP1A1, AKT1, HMOX1, CTNNB1, FYN, MAP2K7, TP53
Transcription factor binding	15	9.22E-09	1.13E-06	JUN, PARP1, CDKN2A, GPX3, STAT3, HMGB1, FOS, HIF1A, ESR1, AR, CCND1, MYC, MAPK1, CTNNB1, TP53
Protein phosphatase binding	9	1.10E-08	1.13E-06	CEACAM1, CDKN1B, ERBB2, STAT3, CTNNB1, MAPK14, MAP2K7, TP53, EGFR
Protein binding	90	3.45E-08	2.41E-06	CDKN1A, FEN1, CDKN1B, SERPINA1, CXCL8, TFRC, OGG1, HILPDA, HMGB1, BRCA2, TNF, IGF1R, SMC2, SOX2, GJA1, CASP8, CCND1, MYC, CHEK2, AKT1, SOX9, TIMP1, HRAS, PDGFRA, EDN1, PARP1, ANXA2, GPX3, MMP2, WNT5A, APOA1, FOS, MMP9, TGFBR2, ACE2, AR, AZGP1, CEACAM1, DPYD, KIT, SERPING1, TLR4, TP53, AVPI1, CRP, CCL11, PCNA, GSTP1, COX17, PDGFB, PTGS2, USP18, HIF1A, EGFR, EFNB2, ERBB3, TERT, PRDX1, CCL5, ERBB2, HMOX1, MAPK1, FYN, MAP2K7, MAPK3, EGR1, JUN, TGFB1, NOS2, CDKN2A, INSR, CAV1, STAT3, FN1, EEF2, MAPK14, ESR1, PRDX6, MTOR, SOD1, VEGFA, IL4, IL6, CYP1A1, PDCD4, CTNNB1, FAS, PTX3, NFE2L2, BCL2L1
Cytokine activity	12	3.53E-08	2.41E-06	IL4, IL6, EDN1, TGFB1, IFNG, IL1B, WNT5A, TIMP1, HMGB1, TNF, IL2, VEGFA

Term	Count	P value	FDR	Genes
Protein homodimerization activity	21	1.14E-07	6.65E-06	PDGFRA, JUN, TGFB1, GSTM1, NOS2, TFRC, PDGFB, TYMS, PTGS2, SOD1, VEGFA, CEACAM1, TERT, ERBB3, CHEK2, CCL5, KIT, DPYD, CAT, HMOX1, BCL2L1

Table 2.
Gene ontology analysis of Cr interacted genes.

Term	Count	P value	FDR	Genes
Pathways in cancer	41	5.78E-24	5.60E-22	CDKN1A, CDKN1B, CXCL8, GSTP1, PDGFB, BRCA2, PTGS2, HIF1A, EGFR, IGF1R, CASP8, FGF9, CCND1, MYC, ERBB2, AKT1, MAPK1, HRAS, MAPK3, PDGFRA, JUN, TGFB1, NOS2, CDKN2A, MMP2, WNT5A, STAT3, FN1, LAMB1, FOS, MMP9, MTOR, TGFBR2, VEGFA, AR, IL6, KIT, CTNNB1, FAS, TP53, BCL2L1
Proteoglycans in cancer	30	1.79E-21	8.69E-20	CDKN1A, HIF1A, TNF, EGFR, IGF1R, ERBB3, CCND1, MYC, ERBB2, AKT1, MAPK1, HRAS, MAPK3, TGFB1, CAV1, MMP2, WNT5A, STAT3, FN1, MIR21, MAPK14, MMP9, ESR1, MTOR, VEGFA, PDCD4, CTNNB1, FAS, TP53, TLR4
HIF-1 signaling pathway	21	3.91E-18	1.26E-16	CDKN1A, EDN1, CDKN1B, NOS2, TFRC, INSR, STAT3, HIF1A, EGFR, MTOR, IGF1R, VEGFA, IL6, IFNG, ERBB2, AKT1, HMOX1, MAPK1, TIMP1, TLR4, MAPK3
Chagas disease (American trypanosomiasis)	21	2.09E-17	5.07E-16	JUN, TGFB1, ACE, CXCL8, NOS2, FOS, MAPK14, TNF, IL2, TGFBR2, IL6, CASP8, IFNG, IL1B, CCL5, FAS, CCL2, AKT1, MAPK1, TLR4, MAPK3
Bladder cancer	14	1.10E-14	2.13E-13	CDKN1A, CXCL8, CDKN2A, MMP2, MMP9, EGFR, VEGFA, CCND1, MYC, ERBB2, MAPK1, HRAS, TP53, MAPK3
Hepatitis B	21	1.87E-14	3.02E-13	CDKN1A, JUN, TGFB1, CDKN1B, PCNA, CXCL8, STAT3, FOS, TNF, MMP9, IL6, CASP8, CCND1, MYC, FAS, AKT1, MAPK1, HRAS, TP53, TLR4, MAPK3
Prostate cancer	16	1.94E-12	2.69E-11	PDGFRA, CDKN1A, CDKN1B, PDGFB, EGFR, MTOR, IGF1R, AR, CCND1, ERBB2, AKT1, MAPK1, CTNNB1, HRAS, TP53, MAPK3
TNF signaling pathway	17	2.75E-12	3.34E-11	JUN, EDN1, FOS, PTGS2, MAPK14, TNF, MMP9, IL6, CASP8, IL1B, CCL5, FAS, CCL2, AKT1, MAPK1, MAP2K7, MAPK3
Pancreatic cancer	14	7.57E-12	8.16E-11	TGFB1, CDKN2A, STAT3, BRCA2, EGFR, TGFBR2, VEGFA, CCND1, ERBB2, AKT1, MAPK1, TP53, BCL2L1, MAPK3
HTLV-I infection	23	1.29E-11	1.25E-10	PDGFRA, EGR1, CDKN1A, JUN, TGFB1, PCNA, CDKN2A, WNT5A, PDGFB, FOS, TNF, IL2, TGFBR2, IL6, TERT, CCND1, CHEK2, MYC, AKT1, CTNNB1, HRAS, TP53, BCL2L1

Table 3.
Pathway analysis for the chromium interacting genes related to Respiratory Disease.

and related networks. The TargetScan library in EnrichR can be used for the possible miRNA interaction. TFs are identified to be significantly associated with the genes involved in the respiratory disease. TRIM28, NFE2L2, EGR1 GATA2, PPARG,

ZMIZ1 and ESR1 are significant for respiratory disease influencing DEGs. The regulated genes for each of these TFs for chromium toxicity are shown in **Table 4** followed by the miRNAs identified for chromium interacting genes involved in the Respiratory diseases in **Figure 1**.

3.3 Comparable chemicals

Information about biological effects of a chemical at genetic level can be extensively extracted from CTD to create new hypotheses with a lot of interaction pathways and networks among genes-contaminants and diseases [22].

This highly contributes in identifying similar contaminants responsible for specific diseases. Comparable chemicals extracted from CTD for the possible sharing with many of the networks common to chromium in respiratory disease are given in **Table 5**. Mercury, SB 203580, 2-(4-morpholinyl)-8-phenyl-4H-1-benzopyran-4-one, 2,3-dimethoxy-1,4-naphthoquinone, were found interacting with 102, 81,77and 61 chromium-iInteracting genes in Respiratory disease.

Term	Overlap	P-value	Adjusted P-value	Combined Score	Gene
TRIM28	9/210	1.02E-05	9.92E-04	82.94725092	EFNB2;EGR1;JUN;PARP1;ERBB3; WNT5A;SOX9;HIF1A;SOD1
NFE2L2	19/1022	3.76E-05	0.001413873	32.51065582	CDKN1A;GSTM1;WNT5A;FN1; PTGS2;ESR1;PRDX6;VEGFA;EFNB2; GJA1;GCLC;PRDX1;DPYD;CAT; CYP1B1;HMOX1;FYN;PTX3;AVPI1
EGR1	10/315	4.37E-05	0.001413873	53.21069483	EGR1;JUN;SERPINA1;STAT3;AKT1; MAPK1;ESR1;MMP9;EGFR;SOD1
GATA2	15/772	1.64E-04	0.003970477	28.45933865	JUN;CDKN1A;EDN1;GPX1;MMP2; LAMB1;FOS;MAPK14;IGF1R;IL4; CHEK2;PDCD4;IDS;IER2;BCL2L1
PPARG	12/535	2.10E-04	0.004075587	31.58677746	EFNB2;PDGFRA;JUN;CDKN1A; CASP8;INSR;HILPDA;CYP1B1;FOS; BCL2L1;SOD1;VEGFA
ZMIZ1	16/914	3.19E-04	0.005162124	23.65905038	EGR1;TGFB1;CDKN1B;GSTP1; MIR21;SOD1;VEGFA;GCLC;MYC; PRDX1;CAT;IDS;MAPK1;CTNNB1; IER2;AVPI1
ESR1	6/154	5.36E-04	0.006683025	48.17527072	EDN1;SERPINA1;STAT3;CYP1B1; FOS;ESR1
CTCF	24/1790	5.51E-04	0.006683025	17.25236108	PDGFRA; EGR1; JUN; EDN1; TGFB1; PCNA; CAV1; APOA1; EEF2; MAPK14; IGF1R; VEGFA; EFNB2; GCLC; IL6; CEACAM1;CASP8; ERBB3;MYC;CYP1B1;MAP2K7;TP53; BCL2L1;NFE2L2
MYC	11/573	0.001387281	0.014951805	20.72253353	GJA1; GCLC; PCNA; TFRC; CCND1; TERT;PARP1;EEF2;TP53;IER2;SOD1
RAD21	17/1265	0.003726145	0.036105828	12.44311845	PDGFRA; JUN; EDN1; PCNA; APOA1; EEF2; MAPK14; SOD2; VEGFA; EFNB2; IL6; CEACAM1; CASP8; MYC; TP53; BCL2L1;NFE2L2

Table 4.
Transcription factors for the chromium interacting genes involved in the Respiratory diseases.

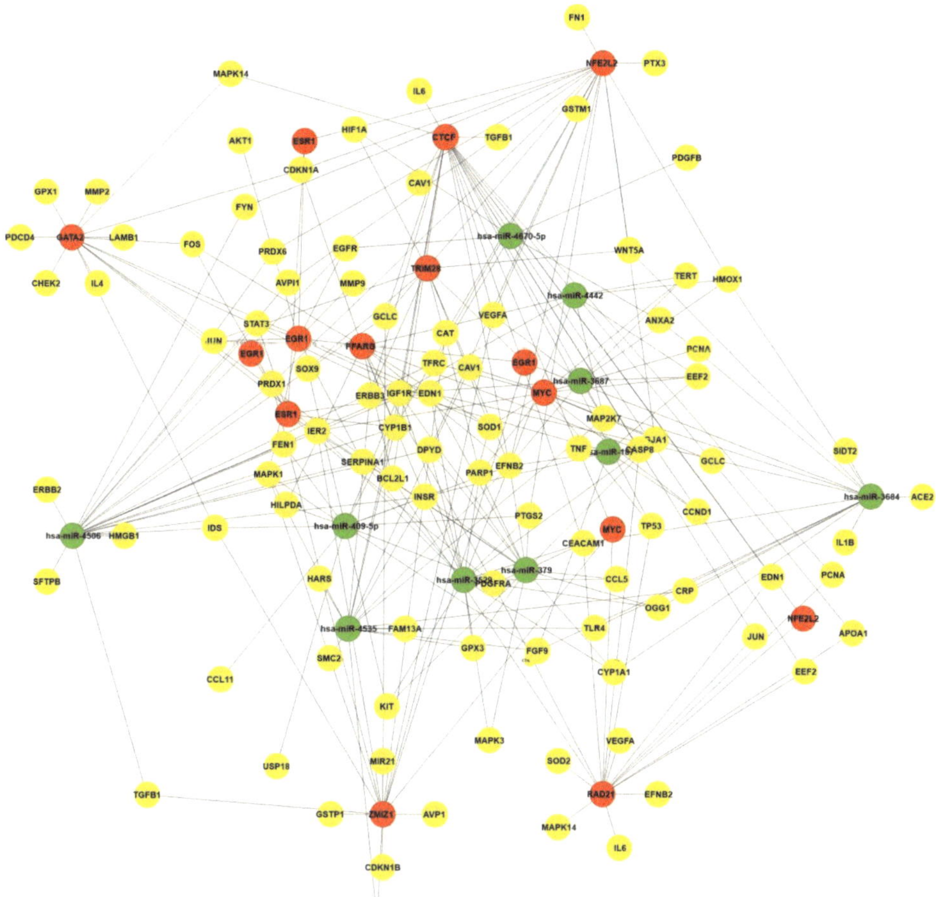

Figure 1.
Gene-TFs-miRNA Interaction Network.

Chemical	CAS RN	Similarity Index	Common Interacting Genes for Respiratory disease
2,3-dimethoxy-1,4-naphthoquinone	6956-96-3	0.174285714	61
Niacin		0.150997151	53
Antimony	7440-36-0	0.143333333	43
Antimony Potassium Tartrate	28300-74-5	0.132183908	46
naringin	10236-47-2	0.131016043	49
SB 203580		0.13022508	81
Mercury	7439-97-6	0.129606099	102
Rutin	153-18-4	0.129518072	43
alpha-Tocopherol	59-02-9	0.126262626	50
cobaltiprotoporphyrin	14325-03-2	0.124338624	47
4-(4-fluorophenyl)-2-(4-hydroxyphenyl)-5-(4-pyridyl) imidazole		0.122615804	45
Luteolin	491-70-3	0.122395833	47

Chemical	CAS RN	Similarity Index	Common Interacting Genes for Respiratory disease
2-(4-morpholinyl)-8-phenyl-4H-1-benzopyran-4-one	154447-36-6	0.122222222	77
Thioctic Acid	62-46-4	0.121890547	49
Cholesterol, Dietary		0.120943953	41
2-(2-amino-3-methoxyphenyl)-4H-1-benzopyran-4-one		0.120385233	75
pyrazolanthrone		0.117117117	65
Docosahexaenoic Acids	25167-62-8	0.117073171	48

Table 5.
Chemicals having comparable sets of interacting genes to chromium.

4. Conclusions

Chromium (VI) is a vital toxic environmental pollutant having various sources including mine tailings. This chapter enlighten respiratory disease accelerated as well as caused due to chromium exposure at genetic level following bioinformatics method that leverages curated data from the public database CTD to generate novel sets of information. This strategy does not require a priori knowledge of the toxicant, biological system, or adverse outcome, and it can be used to identify potential molecular and biological intermediary steps that help fill in knowledge gaps connecting chemical exposures with outcomes for environmentally influenced diseases. With the existed data libraries (mainly CTD, GO, pathway, TFs and miRNA relate databases), bioinformatics web-based tools (David and EnrichR), BPs, CCs, MFs, KEEG signal pathways and gene regulation in the chromium-gene-disease networks were presented. In this study, 127 genes are identified as affected by exposure CR(VI), which are majorly regulated by 10 TFs and 10 very high target miRNAs. The Gene-TFs-miRNAs network recognises maximum interacted genes (EFNB2, IGF1R, CYP1B1, INSR, and VEGFA) and TFs (ZMIZ1, NFE2L2, CTCF and RAD21) and miRNAs (hsa-miR-4506, hsa-miR-379, hsa-miR-3529, hsa-miR-4535, hsa-miR-3684, and hsa-miR-409-5p). The significant biological process (positive regulation of gene expression and positive regulation of nitric oxide biosynthetic process), Cellular Component (extracellular space and protein complex) and Molecular Function (identical protein binding and enzyme binding) are influenced by chromium exposures. From pathway analysis of Cr (VI) influence on respiratory disease, maximum of DEGs are identified to be involved in various pathways in cancer (41 nos.) followed by proteoglycans in cancer (30 nos.), and HTLV-I infection (23 nos.) and so on. Comparable contaminants analysis has recognised Mercury, SB 203580, 2-(4-morpholinyl)-8-phenyl-4H-1-benzopyran-4-one and 2-(2-amino-3-methoxyphenyl)-4H-1-benzopyran-4-one to have maximum common DEGs with Cr (VI) exposure.

Author details

Jyoti Kant Choudhari[1,2], Jyotsna Choubey[2], Mukesh Kumar Verma[1,3], Anand Kumar Jayapal[3] and Biju Prava Sahariah[1*]

1 Chhattisgarh Swami Vivekanand Technical University, Bhilai, C.G, India

2 Raipur Institute of Technology Raipur, C.G, India

3 National Institute of Technology Raipur, C.G, India

*Address all correspondence to: biju.sahariah@gmail.com

References

[1] Bogden, John D.; Klevay, Leslie M., eds. (2000). "Trace Elements and Minerals in the Elderly § Chromium". *Clinical Nutrition of the Essential Trace Elements and Minerals: The Guide for Health Professionals*. Springer Science +Business Media. p. 189. ISBN 978-1-61737-090-8 – via Google Books.

[2] *Barceloux, Donald G.; Barceloux, Donald (1999). "Chromium".* Clinical Toxicology. *37 (2): 173–194.* doi: 10.1081/CLT-100102418. PMID 10382554.

[3] EPA (1998). "Toxicological Review of Trivalent Chromium. CAS No. 16065-83-1. In support of Summary Information on the Integrated Risk Information System (IRIS). U.S. Environmental Protection Agency, Washington, D.C."

[4] PHS (1953). "Health of workers in chromate producing industry: A study." Washington, DC: U.S. Public Health Service. Publication no.192.

[5] Sluis-Cremer, G. and R. du Toit (1968). "Pneumoconiosis in chromite miners in South Africa." Br J Ind Med **25**: 63-67.

[6] Moller, D. R., S. M. Brooks, *et al.* (1986). "Delayed anaphylactoid reaction in a worker exposed to chromium." Journal of Allergy & Clinical Immunology **77**(3): 451-6.

[7] Agency for Toxic Substances and Disease Registry (2000). "Toxicological Profile for Chromium." http://www.atsdr.cdc.gov/toxprofiles/tp.asp?id=62&tid=17.

[8] Lindberg, E. and G. Hedenstierna (1983). "Chrome plating: symptoms, findings in the upper airways, and effects on lung function." Archives of Environmental Health **38**(6): 367-74.

[9] Gibb, H. J., P. S. Lees, *et al.* (2000). "Lung cancer among workers in chromium chemical production. [see comment]." American Journal of Industrial Medicine **38**(2): 115-26.

[10] Novey, H. S., M. Habib, *et al.* (1983). "Asthma and IgE antibodies induced by chromium and nickel salts." Journal of Allergy & Clinical Immunology **72**(4): 407-12.

[11] Langard S (1980). "A survey of respiratory symptoms and lung function in ferrochromium and ferrosillicon workers." Int Arch Occup Environ Health(46): 1-9.

[12] Henderson, R. F., A. H. Rebar, *et al.* (1979). "Early damage indicators in the lungs. IV. Biochemical and cytologic response of the lung to lavage with metal salts." Toxicology & Applied Pharmacology **51**(1): 129-35.

[13] Grondin CJ, Davis AP, Wiegers TC, King BL, Wiegers JA, Reif DM, Hoppin JA, Mattingly CJ. Advancing exposure science through chemical data curation and integration in the comparative toxicogenomics database. Environmental Health Perspectives. 2016; 124:1592–1599. DOI: 10.1289/EHP174

[14] King BL, Davis AP, Rosenstein MC, Wiegers TC, Mattingly CJ. Ranking transitive chemi- cal-disease inferences using local network topology in the comparative toxicogenomics database. PLoS One. 2012;7:e46524. DOI: 10.1371/journal.pone.0046524

[15] Kanehisa M, Goto S. KEGG: kyoto encyclopedia of genes and genomes. Nucleic Acids Res. 2000;28:27–30

[16] Huang DW, Sherman BT, Lempicki RA (2009). Systematic and integrative analysis of large gene lists using DAVID bioinformatics resources. Nat Protoc, 4, 44-57.

[17] Huang da W, Sherman BT, Lempicki RA. Bioinformatics enrichment tools: Paths toward the comprehensive functional analysis of large gene lists. Nucleic Acids Research. 2009;37:1–13. DOI: 10.1093/nar/gkn923

[18] Hobert O. Common logic of transcription factor and microRNA action. Trends in Biochemical Sciences. 2004;29:462–468. DOI: 10.1016/j.tibs.2004.07.001

[19] Hobert O. Gene regulation by transcription factors and microRNAs. Science. 2008;319:1785– 1786. DOI: 10.1126/science.1151651

[20] Chen EY, Tan CM, Kou Y, Duan Q, Wang Z, Meirelles GV, Clark NR, Ma'ayan A. Enrichr: Interactive and collaborative HTML5 gene list enrichment analysis tool. BMC Bioinformatics. 2013;14:128. DOI: 10.1186/1471-2105-14-128

[21] Kuleshov MV, Jones MR, Rouillard AD, Fernandez NF, Duan Q, Wang Z, Koplev S, Jenkins SL, Jagodnik KM, Lachmann A, McDermott MG, Monteiro CD, Gundersen GW, Ma'ayan A. Enrichr: A comprehensive gene set enrichment analysis web server 2016 update. Nucleic Acids Research. 2016;44:W90–97. DOI: 10.1093/nar/gkw377

[22] Davis AP, Murphy CG, Saraceni-Richards CA, Rosenstein MC, Wiegers TC, Hampton TH, Mattingly CJ. GeneComps and ChemComps: A new CTD metric to identify genes and chemicals with shared toxicogenomic profiles. Bioinformation. 2009;4:173–174. DOI: 10.6026/97320630004173.

Section 2

Trace Elements in Environmental Problems

Chapter 6

Trace Elements in Urban Particulate Matters: Variations in Serum Levels, Inhalation Bioaccessibility, Health and Disease Effects

Emmanuel Gbenga Olumayede, B. Babalola and I. Oghenovo

Abstract

Trace elements-bound to particulate matters are often become entrained in human respiratory airway, deposited in human nasal cavity and made available for absorption by human tracheobronchial. It has been assumed that variability and bioaccessibility of elements in the serum correlate with some health and diseases. This chapter is a summary of previous works on bioaccessibility of trace elements bound to inhale particulates using different kinds of simulated body fluids. Presented also are evidences of serum variation in some respiratory diseases, such as chronic obstructive pulmonary disease (with or without hypertension), emphysema, bronchiectasis and bronchial asthma, non-tuberculose mycobacterial (NTM) lung disease, idiopathic pulmonary fibrosis (IPF).

Keywords: trace elements, particulate matters, inhalation bioaccessibility, respiratory fluid, health effect, disease

1. Introduction

Since the industrial revolution, a considerable increase in air pollution has been noted. According to a World Health Organization air quality report [1], inhalation of trace elements bound to airborne particulates is worsening air pollution in cities of the world, thereby causing more than 2 million premature deaths annually. In urban centers, particulate matters are major pollutants in the atmosphere, as they present health risk to dwellers. Urban particulates are known for their heterogeneous mix with diverse natural and anthropogenic origins. The composition can vary depending on geographical location, resuspended soil, atmospheric deposition and sources, which include traffic related particles such as metallic components, eroded road pavement, building construction and demolition, and power generation [2, 3]. The mean daily concentration of PM of $\leq$10 μm in diameter (PM_{10}) ranges from <10 μg/m^3 to 200/m^3 [4]. In 2002 the USEPA reported a range of maximal city concentrations of 25–534 μg/m^3 [5]. These toxic contaminants originated mainly from the anthropogenic emission sources, through ubiquitous applications of elements in urban centers including automobile, industries and domestic fuels combustion [2].

IntechOpen

Quite a lot of researchers have investigated elemental compositions of suspended particulate matters in cities worldwide [4–9]. In most of these studies, elevated levels of trace elements have been observed in atmospheric suspended dust in most cities. For example, Okunola *et al.* [8] reported the presence of Cd, Cr, Ni, Pb, Cu, and Zn in atmospheric settling dust in Kano metropolis of Nigeria. Meanwhile, Mafuyai *et al.* [9] reported that the concentrations of some trace elements were found to be far above the standard limits prescribed by WHO for respirable dust in Jos, Nigeria. Therefore, urban dwellers are exposed to considerable amounts of these elements through inhalation of airborne particulates.

Once inhaled, these particles are deposited in the lung and thereby cause serious health effects. Ruby *et al* [10] reported that more than 80% of the binding mass of particles smaller than 2.5 μm reaches the pulmonary alveoli, where a small fraction is deposited and can stay for months to years. Zwozdziak *et al* [11] has also observed that elements deposition in human respiratory tract decreases with increase depth. Recognizing that dissolution of inhaled particulate-bound metal in the body has been observed to depend on the ability of such metal to be solubilized in body fluids [8], therefore it is only such soluble fraction of the elements which can be taken across the cell membrane through lung pathway that have direct effects on health. Hence, it is important to assess the bioaccessibility of trace elements bound to inhale particles over total metal concentration in particle's matrix.

In this chapter, we aimed to discuss the fates, mechanism of toxicity, and recent trends in assessment of bioaccessibility of trace elements. Attempt was made to understand influence of serum levels on trace elements in some respiratory disorders such as chronic obstructive pulmonary disease (COPD), bronchial asthma. This presentation will not consider routes of exposure other than inhalation of particulate matters.

2. Trace elements

Trace elements are elements present in natural materials at concentration of <1000 $mgkg^{-1}$ [11]. Some of them are essential micronutrients that exist in very low concentrations in the body, forming 0.01% of the total body weight [12] while others are classified as non-essential. Generally, the major trace elements in atmospheric dust are: iron, manganese, zinc, vanadium, chromium, nickel, copper, cobalt, lead, cadmium, mercury.

2.1 The roles of trace elements in biological processes

Some trace elements are essential for human body; for cell metabolism regulation, including activation or inhibition of enzymatic reactions, and regulation of gene and membrane functions.

Many enzymes have trace elements within their structures and these trace elements act as a cofactor to them [13]. These enzymes play important roles in protection of the body by their activatory or inhibitory and antioxidant activities, with defense system molecules in diseases. For example, Iron is an important constituent of succinate dehydrogenase as well as part of heme of the haemoglobin, myoglobin and the cytochromes [14]. Zinc is involved in carbonic acid (Carbonic anhydrase) and in alcohol (alcohol dehydrogenase) formation, and in proteolysis (Carboxypeptidase, leucine, aminopeptidase etc) [15]. Copper is present in many enzymes involved in oxidation (tyrosinase, ceuloplasmin, amino oxidase, cytochrome oxidase) [16]. Changes in the levels of these trace elements decrease the efficiency of the antioxidants systems and lead to hyper-reactivity and inflammation in the respiratory tract [17, 18].

Although, trace elements play important roles in various physiological processes and are crucial for functioning of the immune system. However, excessive accumulation or deficiency of some of these elements in human body may be associated with metabolic disturbance, tissue damage and infectious diseases.

3. Sources of particulate matters and trace elements in urban atmosphere

Human activities have been found to contribute more to environmental pollution due to the everyday manufacturing of goods to meet the demands of the large population [19]. Particulate matters in the environment emanate from two main sources: (i) Environmental sources: this include processes like forest fires, marine water sprays, and volcanic emissions, and (ii) Human-derived sources include a variety of largely industrial sources, like cement and metals manufacturing, incinerators, power plants, refineries, smelters, and vehicular exhaust and dust. Include volcanic products, minerals which occur naturally in the environment Anthropogenic activities such as Oil, natural gas production, petroleum utilization, combustion products (ie, lead in gasoline), manufacturing/industrial wastes and byproducts; commercial products (ie, lead paint in houses), or spills thereof (ie, commercial chemicals), municipal waste incinerators, landfills, sewage sludge disposal etc. **Figure 1** illustrates the cycle of trace elements in atmosphere of urban centers.

Meanwhile, trace elements in the atmosphere originate mainly from anthropogenic emission sources, through ubiquitous applications of elements in urban centers including automobile, industries and domestic fuels combustion [20]. Trace elements emitted in wind-blown dusts are mostly from industrial areas. Some important anthropogenic sources which significantly contribute to the atmospheric pollution in urban centers include automobile exhaust which releases lead; smelting which releases arsenic, copper and zinc; insecticides which release arsenic and burning of fossil fuels which release nickel, vanadium, mercury, selenium and tin. Other metals reported on the particles are iron (Fe), Zinc (Zn), and Nickel (Ni), and recently with the use of the catalytic converters an increase in the presence of Platinum (Pt), Paladium (Pd) and Rhodium (Rh) in the particles inhaled has been observed.

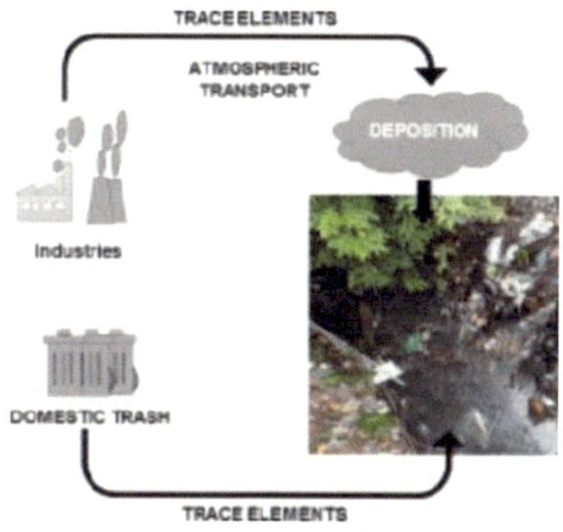

Figure 1.
Cycling of trace elements in the urban atmosphere. Source: http://doi.org/10.1016/j.scitotenv.2019.13447.

4. Routes of exposure and safety limit of some trace elements

For a better understanding of the significances of trace element in human health, it is important to have some knowledge of their routes of exposure. Human are exposed to trace elements in the environment through different routes including ingestion, inhalation of dusts, gases, aerosols and dermal absorption (through skin). The main routes of exposure to trace elements bound to particulate matter (PM) in urban centers include occupational exposure through activities listed below for some specific elements such as:

4.1 Cadmium (Cd)

Cd is an environmentally widespread toxic element. It is classified as a group I carcinogen by IARC (International Agency for Research on Cancer) and has been associated with lung cancer [21]. The modes of human exposure are contamination food, drinking water, occupational or by inhalation in polluted air. Occupational exposure to cadmium primarily takes place in industrial factories such as zinc smelters, battery manufacturing and metal-recovering factories, cadmium-refining companies, production units for paint and pigment. The threshold safety cadmium exposure level has been set at 2.5 μg/kg body weight per week [21]. Cadmium (Cd) exposure is known to induce pulmonary damage such as emphysema and lung cancer [22].

4.2 Lead (Pb)

Worldwide, lead in atmosphere originates from human activities following its uses as; gasoline additive, paints, cosmetics, ceramic glaze, etc. [23]. Lead enters the human body by ingestion or inhalation. According to the WHO-OSHA, the established safety standard for blood lead in workers is 40 μg/dL. However, it has been suggested that the criterion for elevated blood levels in children is too high in adults therefore recommended a new set of guidelines levels >15 μg/dL [24].

4.3 Manganese (Mn)

Atmospheric Manganese originated from gasoline additive, methylcyclopentadienyl manganese tricarbonyl (MMT) is a putative modulator of dopamine biology (the primary target of Mn neurotoxicity) [25].

4.4 Chromium (Cr)

Chromium is widely used in the industry for the production of stainless steel, chromium plating, and spray-painting. According to World Health Organization (WHO) [26], the long term exposure of Cr (VI) levels of over 0.1 ppm causes respiratory problems, liver and kidney damage, and carcinogenicity. According to epidemiological studies, the hexavalent form [Cr (VI)] of this metal, appears to be drastically toxic and carcinogenic, thus it has been classified as carcinogenic to humans by the IARC [27].

4.5 Aluminum (Al)

Aluminum and its compounds [28] are released into the atmosphere during activities such as aluminum mining, processing, production and recovery. The skin, nose, lung and gastrointestinal tract is a route for the uptake of aluminum in the body [29]. Therefore, people close to industrial areas may be exposed to aluminum through inhalation of airborne particulates.

4.6 Arsenic (As)

Elemental arsenic is a metalloid that exists in valency states; trivalent AS^{III}, pentavalent As^{V} in the environment. The main sources of exposure to arsenic include; occupational, environmental and medicinal sources. The safety level of arsenic has been lowered from 50 ppb to 10 ppb by United State Environmental Protection Agency [30]. The presence of arsenic in airborne particulate matter is considered a risk for certain diseases. All the potential pathways of its exposure seem to have adverse effect on human health [31]. Arsenic exposure has been repeatedly associated with lung carcinogenesis [32].

4.7 Vanadium

Vanadium is a major transition element that is released primarily by the burning of fossil fuels, including petroleum, oil, coal, tar, bitumen, and asphaltite. Among Vanadium compounds, Vanadium pentoxide is highly toxic [33]. The IARC classified it as a possible carcinogen to humans (Group 2B) in 2003 [34].

4.8 Zinc

Occupational studies of workers exposed to zinc by inhalation (usually in the presence of other trace elements such as copper, lead, arsenic, and chromium) have not implicated zinc as a risk factor for cancer [35].

5. Behavior, fate, and effects of trace elements in the respiratory tract

The fate and behavior of trace elements in respiratory tract are fundamental to understanding of their health effects and in recent time has become a key aspect of potential health risk assessment.

5.1 The respiratory tract and deposition of PM in the lung

Particulate matters are inhaled during breathing. Upon inhalation, deposition of the particles in the lung may occur through five different mechanisms: sedimentation (gravity), inertial impaction, interception (particle-surface contact), electrostatic deposition, and diffusion. These mechanisms generally occur in different regions of the respiratory tract [36, 37]. Human respiratory tract can be divided into the upper respiratory region (nasal airway, pharynx and larynx), the lower respiratory region (trachea and bronchi) and the alveolar region. **Figure 2a** shows the particle size distribution in human respiratory tract. Meanwhile **Figure 2(b)** llustrates the health risk of trace elements and bioaccessibility questions. The extent of particle deposition in the lung is determined by the physicochemical properties of the particles, such as size, shape, density, and surface chemistry [38] (see **Figure 2a**). Breathing conditions, like ventilation rate, mouth or nose breathing, and airway geometry are other factors that affect particle deposition [39]. The transportation of particles into the lung can be explained by their aerodynamic diameter [40]. Meanwhile, materials with an aerodynamic diameter below 5 μm are predominantly deposited in the alveolar regions of the airways [41].

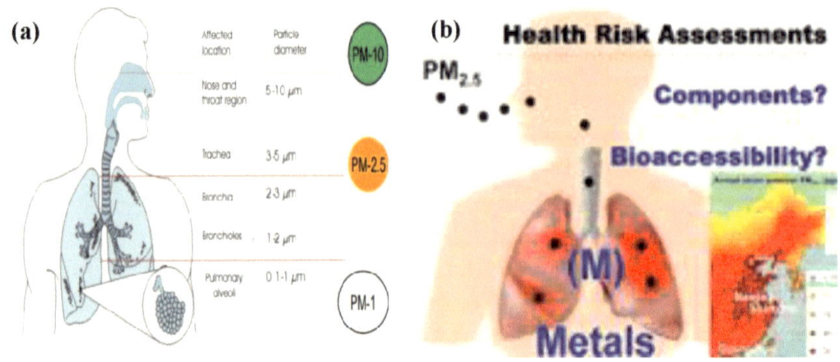

Figure 2.
(a) Dust particle sizes distribution in human respiratory tract (b) human health risk and bioaccessibility questions.

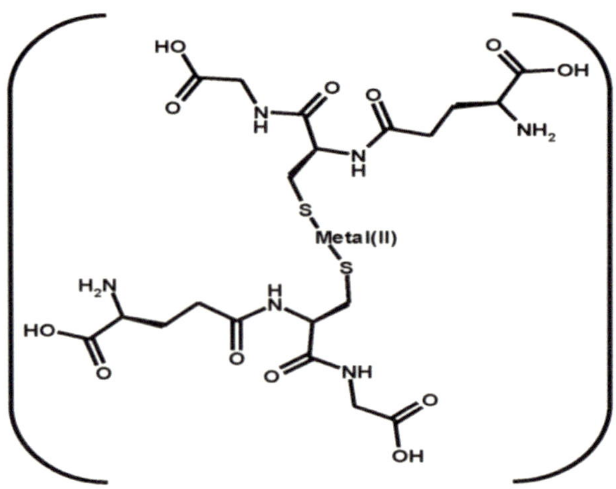

Figure 3.
Glutathione-trace element complex.

When trace elements are absorbed through respiratory tract, it is transported in blood bound to metallothionen [42]. **Figure 3** shows an example of such complex, where they form complex with glutathione. This is then followed by alteration of homeostasis [43], thus directly increasing the oxidative stress and lipid peroxidation.

5.2 Mechanisms of inhaled trace elements toxicity

A primary mechanism for most trace elements toxicity is their effects on cells which has been ascribed to the oxidative stress promoting actions, as observed in *in vivo* [44] and most importantly, the inactivation of enzyme systems by binding to sulfhydryl groups [45] of proteins. The mechanisms of their actions include genetic change reactions; reactive oxygen free radicals and adduct formations, oxidative stress, and inflammation [46].

5.2.1 Reactive oxygen species (ROS) generation

Reactive oxygen species (ROS) such as superoxide, hydroxyl radical, nitric oxide radical are byproducts of metabolic processes. External substances such as smoke, cigerate, pesticides and inhalation of trace elements -bound particulate matters can also cause the formation of free radicals in the body. Trace elements in particulate matters have been reported to cause oxidative stress. For example, pentavalent form of vanadium is reported to cause ROS generation, thus induce oxidative stress, DNA damage, and activation of hypoxia signaling [47]. Oxidation stress is a phenomenon caused by an imbalance between production and accumulation of oxygen reactive species in cell and tissues and the ability of a biological system to detoxify these reactive products [48]. Cadmium causes liver damage mainly by induction ROS inducing lipoperoxidation via Fenton reaction [49]. The increment of ROS induces DNA damage, proteins oxidation and lipid peroxidation. Copper ions are well suited to facilitate formation of ROS that can damage biomolecules, including DNA and chromatin.

5.2.2 DNA adducts formation

The genetic changes reaction of trace elements involves: formation of DNA-protein cross-links, single and double strand DNA breaks [49, 50]. The reaction of elemental ions with nucleic acid lead to a variety of dramatic effects on the nucleic acid structure e.g. crosslinking of polymer strands, degradation to oligomer and monomers, stabilization or destabilization, and the mispairing of bases. For example, Copper can directly bind with high affinity to DNA molecule; this binding can modify the conformational structure of DNA promoting carcinogenesis [51]. Cadmium also produces genotoxicity by the production of DNA single strand breaks and damage and competes for binding at sites (specifically with a zinc finger motifs that are important in gene regulation, enzyme activity, or maintenance of genomic stability [52]).

5.3 Concept of bioavailability and bioaccessibility

In toxicological study, the potential health risks of individual elements bound to inhale particulate matter depend on particle size, inhalability, bioavailability/bioaccessibility, exposure dose and deposition/retention in respiratory tract [53, 54]. Recently, it was emphasized that bio-toxicities of trace metals depend not only on the concentration as expressed by total amount, but also on their geochemical fractions and bioavailability [55]. Bioavailability is the fraction of total elements that can enter the human systemic circulation and exert toxicity on the organs [56]. Meanwhile, bioaccessibility refers to the fraction of contaminant that may become available for absorption *e.g.*, solubilized in the respiratory tract fluid or volatilized into inhaled air and released from the matrix in a topically absorbable form. Bioaccessibility (%) can be defined as the ratio of soluble fraction of trace elements in simulated lung fluids (SLF) to the total concentrations.

5.3.1 Bioaccessibility of trace elements bound to particulate matter

The dissolution of particulate-bound metal in the body has been observed to depend on the ability of such element to be bioaccessible (solubilized) in body fluids after inhalation [57]. Different particulate-bound elemental species behaves differently in human body after inhalation and deposition, depending on their bioaccessibility in lung fluids. In general, high bioaccessible elements are easily taken up by the lung fluids and get introduced to human circulatory system. Recognizing that only soluble fraction of the metals which can be taken across the cell membrane through lung pathway has more direct effects on health. Thus, bioaccessibility of trace elements bound to inhale particles over total metal concentration in particle's matrix is being considered important for assessment of the overall health risk associated with inhalation of particulate matters.

Lung Fluids	BCR-723						
	Trace Element						
	Cd	Cr	Cu	Mn	Ni	Pb	Zn
PBS	<LD	0.8 ± 0.5	4.1 ± 1.5	0.9 ± 0.0	<LD	<LD	6.8 ± 0.8
Gamble's	<LD	0.5 ± 0.3	49.9 ± 5.6	1.7 ± 0.0	0.8 ± 0.0	7.8 ± 0.6	44.6 ± 0.8
ALF	81.4 ± 7.6	8.7 ± 0.0	65.2 ± 3.7	5.5 ± 0.1	24.1 ± 3.7	62.0 ± 3.2	76.8 ± 2.2
	NIST2710						
PBS	44.2 ± 21.2	7.8 ± 0.0	8.3 ± 0.2	28.7 ± 0.4	<LD	0.04 ± 0.00	6.2 ± 0.1
Gamble's	86.0 ± 2.8	<LD	47.6 ± 1.4	40.1 ± 0.7	<LD	7.8 ± 0.4	23.7 ± 0.1
ALF	85.3 ± 8.4	<LD	59.7 ± 1.4	44.3 ± 0.2	<LD	55.0 ± 0.5	35.3 ± 0.1
	NIST 1648						
PBS	24.1 ± 6.2	1.3 ± 0.4	7.3 ± 1.8	16.4 ± 1.4	<LD	<LD	4.3 ± 0.2
Gamble's	45.2 ± 4.0	2.7 ± 1.0	49.9 ± 2.7	29.6 ± 0.2	3.3 ± 1.2	9.1 ± 0.9	43.2 ± 0.2
ALF	65.6 ± 5.5	8.7 ± 0.9	55.0 ± 1.1	46.8 ± 2.6	12.2 ± 4.1	75.9 ± 2.2	66.2 ± 2.3

LD, Limit of detection.

Table 1.
Bioaccessibility (%; mean ± SD; n = 3) values of trace elements in the three lung fluids (adopted from [62]).

Emerging studies [58–60] have shown risk assessment using bioaccessibility presents better understanding of the fate of trace elements upon inhalation by children and adults. However, one of the challenges for environmental toxicologist has been development of fluid with properties similar to human tracheobronchial fluids, so as to enable systematic investigation into bioaccessibility and lung deposition of particles in respiratory tracts [61]. Several fluids have been explored to mimic human respiratory tract fluids in investigation of trace elements bioaccessibility. These range from the traditional Gamble's solution to simulated artificial lung fluids (SALF), which is simply a modification of Gamble's solution.

In one of such previous study, [62] reported that pulmonary bioaccessible fraction of Pb and Cd were relatively high (69 and 74% respectively) when lung stimulating solution (artificial lysosome fluid, ALF) was used to extract fine particles. Similarly, [63–67] reported higher bioaccessibility for Cd (88 ± 6.4% for PM_{10} and 91 ± 6.6% for $PM_{2.5}$) when ALF was used as extraction fluid compared to Gamble's solution. Tang *et al* [64] reported that As, Pb, V and Mn showed higher inhalation bioaccessibility extracted by the artificial lysosomal fluid (ALF); while V, As, Sr. and Cd showed higher inhalation bioaccessibility using the simulated lung fluid (SLF), suggesting differences in elemental inhalation bioaccessibility between ALF and SLF extraction. **Table 1** presents the bioaccessibility values of trace elements in the three lung fluids in different reference materials, as reported by [62]. In general, one of the important factors affecting bioaccessibility of trace elements is the influence of fluid's composition and pH.

6. Variation in serum trace elements levels and induced respiratory tract diseases and health problems

Inhalation exposure to trace elements can have significant health impacts on urban dwellers and nearby workers. Unlike other organs, lungs are directly and continuously exposed to high oxygen concentrations, exogenous oxidants, and pollutants: thus, they have the greatest susceptibility to oxidative stress and pollutant toxicity. The existence of concentration gradient within the lung and inter-individual concentration differences reveals the existence of two groups of elements: (i) homogeneously distributed over the lung e.g. elements Br, Cs, Cu, K, Na, Rb, Se and Zn, and (ii) heterogeneously distributed e.g. elements such as Cd, Co, Cr, Pb, Sb, Sc and V [68].

The enrichment of trace elements in the lung tissue is known to result a number of lung diseases. These diseases have been associated with disturbance of trace elements balance [69]. Here, we discussed recent observations on variation of serum levels in diseases such as chronic obstructive pulmonary disease (with or without hypertension), emphysema, bronchiectasis and bronchial asthma, non-tuberculose mycobacterial (NTM) lung disease, idiopathic pulmonary fibrosis (IPF).

6.1 Chronic obstructive pulmonary disease (COPD)

Many trace elements have activator or inhibitory roles in the antioxidants defensive mechanism in diseases. Recent study [70] showed that serum levels of Co, Cu and Fe were higher in COPD patients with pulmonary hypertension compared to COPD patients without pulmonary hypertension. Similarly, [70] reported that the serum copper (Cu) in COPD patients were higher than the control group.

6.2 Bronchial asthma

Bronchial asthma is a chronic inflammatory disease of the respiratory tract with an unknown etiology where inflammation is often associated with an increase generation of ROS [71]. Several trace elements are known to be capable of causing bronchial asthma, such as nickel (Ni), Chromium (Cr), Cobalt (Co) etc. **Table 2** presents the variations in concentrations of some trace elements (Zn, Cu and Se) in serum of asthmatic, as observed in a study [66]. The results showed higher Cu concentration, and Cu/Zn and lower Cu/Se ratios.

6.3 Idiopathic pulmonary fibrosis (IPF)

Idiopathic pulmonary fibrosis is an interstitial lung disease with poor prognosis and an undefined etiopathogenesis [72] leading rapidly to death. It is the most common lung disease with estimated incidence of 2.8–9.3% per 100,000 per year in Europe and America [73]. Particulate matters bound trace elements deposited in the lung may give rise to more or less marked pulmonary fibrosis, depending on intrinsic properties and amount of the particulate matters. Oxidative stress by trace elements contributes to alveolar injury and fibrosis development in patients. A study [74] reported that IPF patients had significantly increased sputum levels of Cd, Cr, Cu and Pb respect to control. **Table 3** presents the variations in concentrations of some trace elements in serum in patents with NTM, TB and healthy as control, as reported by [74].

Mean ± SD	Control (n = 25)	Patient
Zn (μg/mL)	0.83 (0.14)	0.68(0.09)
Cu (μg/mL)	0.76(0.17)	1.10(0.28)
Se (μg/mL)	0.116 (0.022)	0.0057 (0.024)

Table 2.
Variation of trace elements in serum of asthmatic patients [66, 67].

Element (μg/L)	Patients with NTM (n = 95)	Patient with TB (n = 97)	Healthy control (n = 99)
Co (μg/L)	0.24(0.20–0.35)	0.54(0.22–0.83)	0.23(0.19–0.27)
Cu (μg/L)	109(97–134)	129(111–153)	91(82–102)
Cr (μg/L)	0.23(0.19–0.27)	0.23(0.18–0.27)	0.23(0.19–0.28)
Mn (μg/L)	0.90(0.81–1.07)	0.93(0.71–1.31)	0.92(0.80–1.23)
Se (μg/L)	105(95–116)	108(99–119)	115(105–123)
Zn (μg/L)	94(84–107)	84(75–93)	102(92–116)

Table 3.
Serum levels of trace elements in patents with NTM, TB and healthy [75].

Element (μg/mL)	Patient	Control
Cd (μg/mL)	110	54
Cu (μg/mL)	330	635
Pb (μg/mL)	1217	1444
Mn (μg/mL)	399	522
Se (μg/mL)	1496	1443
Zn (μg/mL)	2515	2699

Table 4.
Serum levels of trace elements in patents with Haemodialysis compare with control [76].

6.4 Non-tuberculose mycobacterial lung diseases (NTM)

Non-tuberculose mycobacterial lung diseases are emerging cause of pulmonary infection and are becoming more common in the clinical setting. A recent study [75] showed that serum concentration of copper and molybdenium (**Table 4**) were higher in patients with NTM lung disease (109 vs. 91 μg/dL, $p < 0.001$ and 1.70 vs. 0.96 μg/L, $p < 0.001$). In contrast, the media serum concentrations of Selenium and Zinc were significantly lower in patients with non-tuberculose mycobacterial lung diseases than in healthy control (105 vs. 115 μg/L, $p < 0.001$ and 94 vs. 102 μg/dL, $p < 0.001$).

6.5 Haemodialysis

Oxidants-antioxidants balance is essential for the normal lung function. Both, an increased oxidant and/or decrease antioxidant may reverse the physiologic oxidants-antioxidants balance, leading to lung injury. Available data (**Table 4**) suggested that the levels of Cd, Cr, Pb, and V were higher and the levels of Se, Zn and Mn were lower in hemodialysis patients compare with controls [76].

6.6 Parkinson disease

Parkinson disease, also known as manganism is an extrapyramidal neurological disease characterized by rigidity action tremor, bradykinesia, memory and cognitive dysfunction that occurs in workers exposed to airborne Mn. The element (Mn) in blood crosses the blood brain barrier and accumulates inside the neuron disrupting the synaptic transmission and inducing glial activation [77].

7. Conclusion

Trace elements bound to particulate matter could be trapped and deposited along the nasal cavity through inhalation of air-borne particulate matter. In this chapter, we attempted to understand influence of serum levels and bioaccessibility of trace elements in some respiratory fluids. Our investigation provides evidence that enrichment of trace elements in the lung tissue is known to result a number of lung diseases, such as chronic obstructive pulmonary disease (with or without hypertension), bronchial asthma, non-tuberculose mycobacterial (NTM) lung disease, and idiopathic pulmonary fibrosis (IPF). The findings suggest that serum Cu were higher in asthmatic patients and COPD patients than the healthy. Meanwhile, the levels of Se, Zn and Mn were lower in hemodialysis patients and non-tuberculose mycobacterial lung diseases than in healthy control.

Author details

Emmanuel Gbenga Olumayede*, B. Babalola and I. Oghenovo
Department of Industrial Chemistry, Federal University Oye-Ekiti,
Ekiti State, Nigeria

*Address all correspondence to: emmanuel.olumayede@fuoye.edu.ng

References

[1] WHO (World Health Organization), 2014. Ambient Air Quality and Health, hppp://www.who.int/mediacentre/factsheets/fs313(access).

[2] Kelly FJ, Fussell JC. Size, source and chemical composition as determinants of toxicity attributable to ambient particular matter. Atmospheric Environment, 2012; 60, 504-526.

[3] Grobéty B, Gieré R, Dietze V, Stille P. Airborne particles in the urban environment. Elements, 2010; 6: 229-234

[4] Amato F, Pandolfi M, Moreno T, Furger M, Pey J, Alastuey A, Bukowiecki N, Prevot ASH, Baltensperger U, Querol X. Sources and variability of inhalable road dust particles in three European cities. Atmos. Environ. 2011; *45*: 6777-6787

[5] WHO Exposure to ambient air pollution [internet] WHO.2016 [cited 2016 Dec 29]. Available from http://www.who.int/gho/phe/outdoor-air-pollution/exposure/en/

[6] Celik I, Gallicchio L, Boyd K, Lam TK, Matanoski, G, Tao X, et al. Arsenic in drinking water and lung cancer: a systemic review. Environ. Res. 2008; 108: 48-55

[7] Li F, Zhang J, Huang J, Huang D, Yang J, Song Y, Zeng G. Heavy metals in road dust from Xiandao District, Changsha City, China: Characteristics, health risk assessment, and integrated source identification. Environ. Sci. Pollut. Res., 2016; *23*: 13100-13113.

[8] Obioh IB, Ezeh GC, Abiye OE, Alpha A, Ojo EO, Ganiyu AK. Atmospheric particulate matter in Nigerian megacities, Toxicol Environ Chem. 2013; 95 (3): 379-385.

[9] Okunola OJ, Uzairu A, Uba S, et al. Distribution Pattern of Metals in Atmospheric Settling Dust along Roads in Kano Metropolis, Nigeria. Journal of Applied Chem., 2015, Article ID 739325, 12Pages

[10] Mafuyai GM, Eneji IS, Sha'Ato R. Concentration of Heavy Metals in respiratory Dust in Jos Metropolitan Area, Nigeria. Open Journal of Air Pollution, 2014; 3: 10-19 http://dx.doi.org/10.4236/ajop.2014.31002.

[11] Ruby MV, Lowney YW, Bunge AL, Roberts SM, Gomez-Eyles JL, Ghosh U, Kissel JC, Tomlin-son P, Menzie C. Oral Bioavailability, Solubility or dissolution, and Dermal Absorption of PAHs from Soil-State of the Science. Environ. Sci. Technol., 2016; 50: 2151-2164

[12] Zwozdzaik A, Gini MI, Samek L, Rogula-Kozlowka W, Sowka L, Eleftheriadis K. Implications of the aerosols size distribution modal structure of trace and major elements on human exposure, inhaled dose and relevance to the PM2.5 and PM10 metrics in European pollution hotspot urban area. J. Aerosol Science. 2017; 103: 38-52

[13] Foster WM, Langenback E, Bergofsky EH. Measurement of tracheal and bronchial mucus velocities in man: relation to lung clearance. J Appl Physiol., 1980; 48: 965-971.

[14] Stöber WKW. A simple pulmonary retention model accounting for dissolution and macrophage-mediated removal of deposited polydisperse particles, Inhal. Toxicol., 2001; 13: 129-148.

[15] Sharareh Dehghani. Farid Moore. Luba Vasiluk. Beverley A. Hale The influence of physicochemical parameters on bioaccessibility-adjusted hazard quotients for copper, lead and zinc in different grain size fractions of urban street dusts and soils. Environ

Geochem Health DOI 10.1007/s10653-017-9994-6

[16] Forbes B, O'Lone R, Allen PP, Cahn A,Clarke C, Collinge M, et al. Challenges for inhaled drug discovery and development: Induced alveolar macrophage responses. Adv. Drug Deliv. Rev. Elsevier B.V. 2014; 71:15-33.

[17] Guha Mazumder DN. Chronic arsenic toxicity: clinical features, epidemiology, and treatment: experience in West Bengal, J. Environ. Sci. Health, Part A. 2003; 38 (1): 141e163.

[18] Guha Mazumder DN. Chronic arsenic toxicity & human health, Indian J. Med. Res. 2008; 128: 436e447

[19] Chung JY, Yu SD, Hong YS, Environmental Source of Arsenic Exposure, **Journal of Preventive Medicine & Public Health** (2014); 253-256.

[20] Christopher E. Environmental Science Processes & Impacts, 2013; 15(10) : 1785-1970.

[21] Nieboer E, Gibson BL, Oxman AD, et al. Health effects of aluminum: A critical review with emphasis on aluminum in drinking water. Environ Rev. 1995; 3(1): 29-81.

[22] IARC Working Group on the Evaluation of Carcinogenic Risks to Humans, Cadmium and cadmium compounds, IARC Monogr. Eval. Carcinog. Risks Hum. 1993; 58:119-237.

[23] Jomova K, Valko M. Advances in metal-induced oxidative stress and human disease. Toxicology 2011; 283 (2-3): 65-87.

[24] Kruszewski M. The role of labile iron pool in cardiovascular diseases. Acta Biochim Pol. 2004; 51(2): 471-480.

[25] CONTAM, Scientific opinion of the panel on contaminants in food chain on a request from European commission on Cadmium in food, The EFSA Journal, 2009; 980: 1-139,

[26] Markowitz M. "Lead poisoning". Pediatrics in Review, 2000; 21(10):327-332,

[27] CDC (Centers for Disease Control and Prevention). Preventing lead poisoning in young children: A statement by the for Disease Control, October 1991, US Department of Health and Human Services, Atlanta, Ga, USA,1991.

[28] Soliman MM, Baiomy AA, Yassin MH. Molecular and histopathological study on the ameliorative effects of curcumin against lead acetate-induced hepatotoxicity and nephrototoxicity in Wistar rats, Biol. Trace Elem. Res. 2015; 167:91-102, http://dx.doi.org/10.1007/s12011-015-0280-0.

[29] Styner L, Smith R, Thun M, Schnorr T, Lemen RA. Dose-response and qualitative assessment of lung cancer risk and occupational cadmium exposure. Ann Epidemiol. 1992; 2: 177-194

[30] Shotyk W, Le Roux G. Biogeochemistry and cycling of lead. Met. Ions Biol. Syst. 2005; 43:239-275 (Accessed 25 April 2017), http://www.ncbi.nlm.nih.gov/pubmed/16370121 .

[31] USEPA, 1994. Methods for Derivation of Lung Reference Concentrations and Application of Lung Dosimetery. Environmental Criteria and Assessment, Office Of Health and Environmental and assessment, Office Of Research and Development , USEPA Research Triangle Park, North Caroline EPA/600/8-90/066F.

[32] Oberdorster G, Oberdorster E, Oberdorster J. *Nanotoxicology: an*

emerging discipline evolving from studies of ultrafine particles. Environ Health Perspect, 2005; **113**(7): 823-839.

[33] WHO (World Health Organization), Environmntal Health Criteria 18: Arsenic, World Health Organization, Geneva, Switzerland, 1981; 43-102.

[34] International Agency for Research on Cancer (IARC) Monographs on the evaluation of carcinogenic risks to humans: (1993) vols. 1-58 IARC, Lyon, pp1971-1993.

[35] OECD. Detail review paper on cell transformation assays for detection of chemical carcinogens. DRP No. 31. Fourth draft version.

[36] Brook RD, Fraklin B, Cascio W, Hong Y, Howard G, Lipsett M, et al. Air pollution and cardiovascular disease a statement for healthcare professionals from expert panel on population and prevention science of the America Heart Association. Circulation 2004; 109: 2655-2671.

[37] Andujar, P., et al., *Respiratory effects of manufactured nanoparticles.* Rev Mal Respir, 2011; 28(8): p. e66–e75.

[38] Mossman, B.T., et al., *Pulmonary endpoints (lung carcinomas and asbestosis) following inhalation exposure to asbestos.* J Toxicol Environ Health B Crit Rev, 2011; 14(1-4): 76-121.

[39] Braakhuis HM, et al., *Physicochemical characteristics of nanomaterials that affect pulmonary inflammation.* Part. Fibre Toxicol, 2014; 11: 18.

[40] Jud C, et al., *Nanomaterials and the human lung: what is known and what must be deciphered to realise their potential advantages?* Swiss Med. Weekly, 2013; 143: 13758.

[41] Jones RM, Neef N. Interpretation and prediction of inhaled drug particle accumulation in the lung and its associated toxicity. Xenobiotica. 2012;42:86-93.

[42] Geiser M, Kreyling WG. *Deposition and biokinetics of inhaled nanoparticles.* Part Fibre Toxicol., 2010; 7: 2.

[43] Heyder J, Svarten, MU. Basic principles of particle behavior in the human respiratory tract, In: Bisgaard H, O'Callaghan C, Smaldone GC, Eds. Drug Delivery to the Lungs. Lung Biology in Health and Disease, Marcel Dekker; New York: 2002. p. 21-45.

[44] Kagi JHR, Kogima Y., eds. Chemistry and Biochemistry of metallothionein. Boston: Birkhäuser, 1987; 25-61.

[45] Slater TF. Free-radical mechanisms in tissue injury, Biochem Journal 1984; 222: 1-15,

[46] Joesten MD, Johnson DO, Netterville JT, Wood JL. World of Chemistry, Brooks/cole. Pacific Grove, CA, USA, 1ST Edition, 1990

[47] Vako M, Morris H, Cronin MT. Metals, toxicity and oxidative stress, Current Medicinal Chemistry, 2005; 12 (10): 1161-1208.

[48] Kagi JHR, Kogima Y. Chemistry and biochemistry of metallothionein. Boston: Birkhäuser, eds. 1987. pp. 25-61

[49] Halliwell B, Gutterridge, JMC. Free radical and antioxidant protection: Mechanism and significant in toxicity and diseases. Human Toxicity, 1988; 7: 7-13.

[50] Aust SD, Morehouse LA, Thomas CE. The role of metals in oxygen radical reactions, J. Free Rad. Biol. Med. 1985; 1: 3-25

[51] Eichhorn GL, Butzow JJ, Shin YA. Some effects of metal ions on DNA structure and genetic information

transfer. Proc. Int. Symp. Biomol., Struct., Interactions, Suppl. J. Biosci. 1985; 8 (3&4): 527-535.

[52] Morris DL. DNA-bound metal ions: recent developments. BioMol Concepts 2014; 5(5): 397-407

[53] Eichhorn, GL. In advances in Inorganic Biochemistry (eds Eichhorn, GL and Marzilli) (New York, Elsevier) Vol. 3. P2

[54] Ruby MV, Schoof R, Brattin W, Goldade M, Post G, Harnois M, Mosby DE, Casteel SW, Berti W, Carpenter M, et al. Advances in evaluating the oral bioavailability of inorganics in soil for use in human health risk assessment. Environ. Sci. Technol., 1999; 33: 3697-3705.

[55] Ng JC, Juhasz A, Smith E, Naidu R. Assessing the bioavailability and bioaccessibility of metals and metalloids, Environ. Sci. Pollut. Res. 2015; 22 : 8802-8825

[56] Huang X, Betha R, Tan LY, Balasubramanian R. Risk assessment of bioaccessible trace elements in smoke haze aerosols versus urban aerosols using simulated lung fluids. Atmos. Environ. 2016; 125: 505-511

[57] Cui X, Xiang P, He R, Juhasz A, Ma L. Advances in in vitro methods to evaluate oral bioaccessibility of PAHs and PBDEs in environmental matrices. Chemosphere 2016; 150, 378-389

[58] Caboche J, Esperanza P, Bruno M, Alleman LY. Development of an in vitro method to estimate lung bioaccessibility of metals from atmospheric particles. Journ. Environ. Monit. 2011; 13: 621-630

[59] Boisa N, Elom N, Dean JR, Deary ME, Bird G, Entwistle JA. Development and application of an inhalation bioaccessibility method (IBM) for lead in the PM10 size fraction of soil. Environ. Inter., 2014; 70: 132-142. Doi: 10.1016/j.envint.2014.05.021.

[60] Luo X.-san, Yu S, Li X.-dong. The mobility, bioavailability, and human bioaccessibility of trace elements in urban soils of Hong Kong, Applied Geochemistry, 2012; 27 (5): 995-1004.

[61] Dean JR, Elom NI, Entwistle JA. Use of stimulated epithelium lung fluid in assessing the human health risk of Pb in urban street dust, Sci. of the Total Environment, 2016; http://dx.doi.org/10.1016/j.sciotenv.2016.11.085

[62] Olumayede EG, Oguntimehin I, Babalola B, Ojiodu C, Akinyeye RO, Sodipe OG, Uche J, Ojo A. Development of tracheobronchial fluid for in vitro bioaccessibility assessment of particulates-bound trace elements. MethodsX, 2019; (6): 1944-1949. https://doi.org/10.1016/j.mex.2019.07.027

[63] Pelfrene A, Cave MR, Wragg J, Douay F. In vitro investigations of human bioaccessibility from reference materials using simulated lung fluids, International Journal of Environmental Research and Public Health, 2017; 14: 112.

[64] Fernánndez Espínosa JA, Ternero Rodríguez M, Barragán de la Rosa FJ, Jiménez Sánchez, JC. Atmos. Environ. 2002; 36: 773-780

[65] Tang Z, Hu X, Chen Y, Qiao J, Lian H. Assessment if in vitro inhalation bioaccessibility of airborne particle-bound potentially toxic elements collected using quartz and PTFE filter. Atmospheric Environment 2019; 196: 118-124.

[66] Wiseman CLS, Zereini F. Characterizing metal (loid) solubility in airborne PM10, PM2.5, and PM1 in Frankfurt, Germany using stimulated lung fluid. Atmos. Environ.

2014; 89: 282-289 doi:10.1016/j.atmosenv.2014.02.055

[67] Alpofead JAH, Davidson CM, Littlejohn D. A novel two-step sequential solubility or dissolution tests for potentially toxic elements in inhalable particulate matter transported into the gastrointestinal tract by mucociliary clearance. Anal Bioanal Chem. 2017; 409: 3165-3174

[68] Guo CH, Liu P, Hsia S, Chuang C, Chen P. Role of certain trace minerals in oxidative stress, imflammation, CD4/CD8 lymphocyte ratios and lung function in asthmatic patients. Ann. Clin. Biochem. 2011; 48: 344-351

[69] Ermis B, Armutcu F, Gurel A, Kar L, Demircan N, Altin R, Daemirel E. Trace elements status in children with bronchial asthma. Eur. J Gen Med. 2004; 1(1): 4-8.

[70] Niu J, Liberda EN, Qu S, Guo X, Li X, Zhang J, et al. The role of metal components in the cardiovascular effects of $PM_{2.5}$, PLoS One, 2013; 8(12): e83782.

[71] Sarkar S, Yadav P, Trivedi R, Bansal AK, Bhatnagar D. Cadmium-induced lipid peroxidation and the status of the antioxidant system in rat tissue. Journal of Trace Elements in Medicine and Biology, 1995; 9(3): 144-149

[72] Kirkham P, Rahman I. Oxidative stress in asthma and COPD: antioxidants as a therapeutic strategy. Pharmacology & Therapeutics, 2006; 111: 476-494.

[73] Bargagli E, Lavorini F, P Istolesi M, Rosi E, Prasse A, Rota E, Voltolini L. Trace elements in fluids lining the respiratory system of patients with idiopathic pulmonary fibrosis and diffuse lung diseases. Journal of trace Element Medicine and Biology, 2017; 42: 39-44. Doi.org/10.1016/j.jtemb.2017.04.001.

[74] Hutchinson J, Fogarty A, Hubbard R, McKeever T. Global incidence and mortality of idiopathic pulmonary fibrosis: a systematic review. Eur Respir J. 2015; 46(3):795-806

[75] Forte G, Bocco B, Pisano A, et al. The levels of trace elements in sputum as biomarkers for idiopathic pulmonary fibrosis, Chemosphere 2021; 271(1): 129514. Doi:10.1016/j.chemosphere.2020.129514.

[76] Oh J, Shin SH, Choi R, Kim S, Park HD, Kim SY, Han SA, Koh WJ, Lee SY. Assessment of 7 trace elements in serum of patients with non-tuberculous mycobacterial lung disease, J Trace Element Med Biol. 2019 May; 53:84-90

[77] Tonelli M, Wiebe N, Hemmelgarn, B et al. Trace elements in haemodailysis patient: a systematic review and meta-analysis, BMC Med 2009; 7: 29 htpps://doi.org/10.1186/171-7015-7-25.

Chapter 7

Analysis of *Aristolochia longa* L. Medicinal Plant from Algeria

Zohra Lamari and Houria Negache

Abstract

In recent time, the therapeutic use of medicinal plants has increased all over the world. The efficacy of herbs for curative purposes is often accounted of its mineral and organic constituents. Neutron activation analysis (INAA) has been applied to mineral determination of Aristolochia Longa (bereztem), medicinal plant used to cure some diseases observed in Algeria especially cancer. In this work the mass fractions of Cr (15.22 ± 3.5 μg/g), Na (269.98 ± 25.01 μg/g), La (0.478 ± 0.041 μg/g), K (1.33 ± 0.23 μg/g), Br (1.2 ± 0.19 μg/g), As (0.697 ± 0.038) and Sb (66.09 ± 11.24 μg/g), were determined. This herb was collected from Taourirt Aden Berber village situated in Northern Algeria. Five elements were quantified in certified AIEA standards IAEA-V10 and IAEA-SL1 for checking the accuracy of our procedure. It was noteworthy the values obtained from this work are in good agreement with the certified values, the Z-score values for all elements were $|Z| < 3$. We believe that herb is natural and harmless compared with chemical drugs. Unfortunately the potential toxicity due to the Aristolochia Acids content has required the analysis of Aristolochia Longa by CG/MS and HPLC to highlight this compound. The standard of Aristolochic Acid (Sigma A5512-25 mg Yellow powder lot # wxbb6331V^{PCODE}) was used as reference.

Keywords: Aristolochia Longa, INAA, organic compounds, public health, Algeria

1. Introduction

Bereztem is the common name of Aristolochia Longa, this herb with a delicate aromatic odor is wrongly used as medicinal plant in Algeria [1, 2] and other countries. We find these plants in the temperate and tropical regions [3]. Since antiquity various *Aristolochia* and *Asarum* species have been used in herbal medicines in obstetrics and in treatment of Intestinal affections, coetaneous diseases, wounds, heart palpitation or snakebite, festering wounds, and tumors [4, 5], indeed 60% of drugs approved for cancer treatment are of natural origin [6, 7]. B. Benarba report the cytotoxic and apoptogenic activities of an aqueous extract of *A. longa* in the Burkitt's lymphoma BL41 cell line [8]. Considered also as antidote against some poisonings [9]. It was reported [10, 12] that plants remain in use today especially in the Chinese medicine. All parts of the plant are used in herbal preparations, and aristolochic acids are present in the roots, stems, leaves, and fruit [11, 12]. Exposure to this acid could potentially occur through the ingestion or skin contact to treat wounds, note no published studies of skin absorption of aristolochic acids in humans or experimental animals were found. However certain plant as canadense leaves cause the dermatitis [13]. The dried rhizome of Aristolochia Longa is the one of the part frequently used often without other ingredients. Several reports indicate the use of complementary and alternative medicine (CAM) and a lot of people in Algeria believe that herb is natural and harmless compared with chemical drugs. Unfortunately, they are unaware of its adverse biological effects. The Aristolochic Acids content, alkaloid components are known to be mutagenic and carcinogenic [14]. The clinical syndrome to the Chinese plants Nephropathy (NCP) was reported in first in Belgium for the women having followed a Chinese slimming diet in 1992 after consumption of herbal weight loss preparations containing Aristolochia Fangehi by inadvertence instead Stephania Tetrandra [15]. Kupchan and Doskovitch 1962 [16], have tested the antitumor effects of aristolochic acids in mice and in clinical trials. But when Jackson et al. [17] showed the neph-rotoxicity of aristolochic acid the trials were discontinued. Mix et al. [18]; Kumar et al. [19] Described twelve Aristolochic Acid analogues, the major compounds of AA_S include AAI and its demethoxylates derivative, AAII, generally the levels of AAI are higher than AAII, the **Figure 1** shown their structures. The metabolites are excreted in the urine and the feces. Reported half-lives in New Zealand White rabbits for aristolochic acids I and II were 0.12 hours and 0.27 hours, respectively. Studies in rats show that the metabolites of aristolochic acid I are excreted within 24 hours, whereas metabolites of aristolochic acid II are still present in the urine at 72 hours. Furthermore the curative properties of this plant are based only on traditional knowledge and in our country there are no procedures and regulations applicable about the use and marketing to the healing plants. Regrettably Aristolochia Longa is easily obtained from local markets. Unlike this herb is forbidden in several countries USA [20], Canada, Taiwan, France, and Belgium. The European Commission (EC) (2000) has prohibited aristolochic acid and its salts, as well as Aristolochia species, and their preparations in cosmetic products. The trace element present in *Aristolochia longa* L. rhizomes; determined by Neutron activation analysis (INAA) can may be explain some therapeutic activities. At the same time two (02) methods have been used for the identification of aristololochic acids (AA_S). We made Gas chromatography – Mass spectrometry (CG/MS) and High performance Liquid chromatography (HPLC). This work can constitute a position paper to better use of this natural product by the cancer patients who take this herb.

AAs I (AAI) and II (AAII) (EMEA 2000).

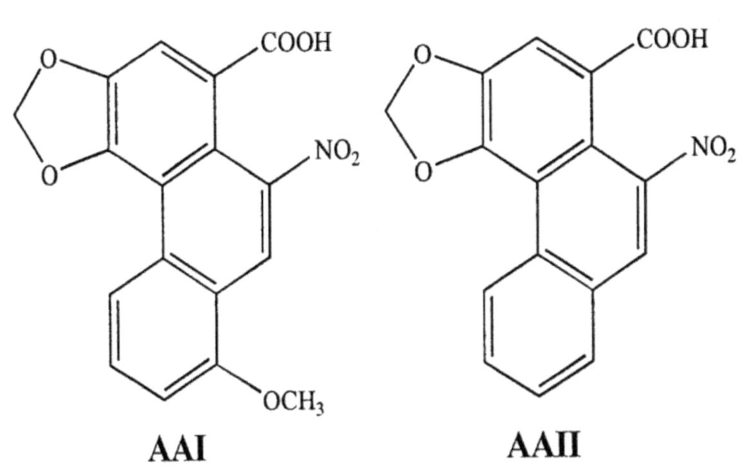

Figure 1.
Chemical structure of nitrophenanthrene carboxylic acids.

2. Experimental

In order to analyze Aristolochia longa L., Rhizomes were collected in May 2016 from Taourirt Aden, Algerian village situated in Kabylia region at 120 km south of Algiers city, Algeria. The vegetal material was washed extensively in distillated water as to remove superficial dust. And then dried at room temperature for one week-end. The dry and hard form of these roots was ground in an electric laboratory blender. One part of the fine powder obtained was prepared for neutron Activation Analysis. The triplicate samples were packed in polyethylene thin target and irradiated during four hours (04) in NUR Algerian reactor, 1 MW research reactor. After 04 and 08 days of delay time the acquisitions were done for determination of Cr, Na, La, K, Br, As and Sb. The analysis was done using HpGe – Canberra detector and the elemental concentrations were performed by Calcon software. The count time was twenty (20) hours for medicinal plants and about five (05) hours for the standards. The certified reference material SDM-2TM Lake (sediment marine lyophilized) was used for calibration. The quality control was done using IAEA biological material V10 (hay powder) and AIEA non biological material SL-1 (lake sediment). Another part of this powder herb was used for the identification of Aristolochic Acids by CG/MS; and HPLC techniques. Many extracted methods have been reported in literature [21]. The extract of AA was prepared by adding 10 ml of methanol to 10 g of *A. longa* dry rhizomes powder for the first extraction and 20 ml of light petroleum for the second extraction, solvent extraction the most commonly used for extraction of AAs [22]. After 24 h of maceration under magnetic stirring at room temperature, the mixture was centrifuged, filtered and then concentrated in a rotary vacuum evaporator (The number of siphon age is ten) indeed the quantification of AA in extract products is less complicated as compared to herbal preparations. However to avoid the loss of chemical information's the extraction method should be no selective to explain the therapeutic aspect. The extracted material was analyzed by CG/MS and HPLC in order to detect the potential presence of AAs in Aristolochia Longa. The standard of Aristolochic Acid recently acquired (Sigma A5512–25 mg Yellow powder lot # wxbb6331V^{PCODE}.) was used as reference. It should be noted that the Aristolochic acids compounds are produced commercially only as reference standards and as research chemicals [12, 23].

3. Results and discussion

Table 1 show the Algerian medicinal plant studied with the botanical name, common name and the part used for treatment. Five elements were quantified in certified AIEA standards IAEA-V10 and IAEA-SL1 for checking the accuracy of our procedure. The values obtained were showed in **Table 2** with the reported certified values. It was noteworthy the values obtained from this work are in good agreement with the certified values; The **Figure 2** presents graphically the plot of Z-score for our elements.

Common name	Botanical name	Family	Part used for treatments
Berztem	Aristoloshia Longa	Aristolochiaceae	Root, aerial part

Table 1.
Botanical name and parts used of the medicinal plant studied.

Element	IAEA-V10		IAEA-SL1	
	Measured	Reported	Measured	Reported
K*	24.65 ± 2.96	21 ± 2		
Cr	6.42 ± 0.798	6.5 ± 0.75	94.92 ± 8.54	104 ± 9
Na*	0.692 ± 0.064	0.5 ± 0.3		
Br	10.82 ± 1.73	8 ± 2	6.82 ± 1.73	9.79 ± 1.58
Sb				

Values given in mg/g.

Table 2.
Quality Control assessment results (μg/g) for the AIEA - certified reference material Samples.

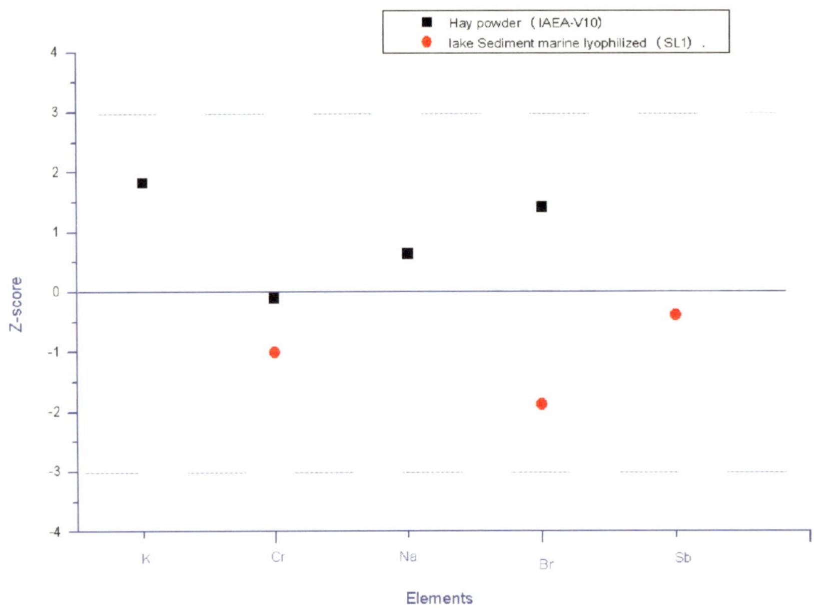

Figure 2.
Z-score values for the elements determined in IAEA standards V10 and SL1.

Table 3 report the elemental concentrations obtained from this work and shows the usefulness of using INAA for the elemental determination, the La, K, Br, As are present at trace levels and the Cr, Sb was found at minor level and Na at the major level. Owing to its high toxicity, the identification of AAs is obviously important indeed these acids cause Aristolochic Acids Nephropathy (NAA), the chronic renal failure according to Grollman et al. [24] and urotherial carcinomas [25, 26].

A new American study reveals that these AAs are more carcinogen than the tobacco [27]. The toxicity of the aristolochic acids has been studied and reported by Mengs and Stotzem [28]. When The Arirstolochic Acids extracted from a medicinal plant are traditionally used in China to cure some diseases, the arthritis and the other inflammations. It has been shown by various authors that these Aristolochic acids have directly toxic on the human gene TP53 (gene suppressor of cancer) [29]. No mutations were identified in rats with chronic renal failure not exposed to aristolochic acids. Similar findings have been reported in humans [30]. The complexity of herbal nomenclature systems used in traditional Chinese medicines may have contributed to the potential exposure to aristolochic acids. As well as the similarity of the Chinese names for *Aristolochia* species and other innocuous herbs can increase risk of inadvertent exposures to aristolochic acids [31].

Even more similar Japanese and Chinese names refer to different plants in Japan and China [32], explained the outbreak of Chinese herb nephropathy in Japan by the substitution of that plant species in Japanese preparations of Chinese herbal medicines. In this study we are analysed the rhizome of Aristolochia Longa to identify the aristolochic acid much to our surprise this organic compound is not reveled in our herb. The **Figures 3** and **4** shows the results obtained by CG/MS and HPLC, the characteristic retention times of AAI Acid present in our reference standard are 42.88 min and 21.758 min relative to the analysis by CG/MS and HPLC respectively.

Element	Aristoloshia Longa
Cr	15.22 ± 3.5
Na	269.98 ± 25.01
La	0.478 ± 0.041
K	1.33 ± 0.23
Br	1.2 ± 0.19
As	0.697 ± 0.038
Sb	66.09 ± 11.24

Table 3.
Elemental Concentrations (μg/g) for Aristoloshia Longa rhizome.

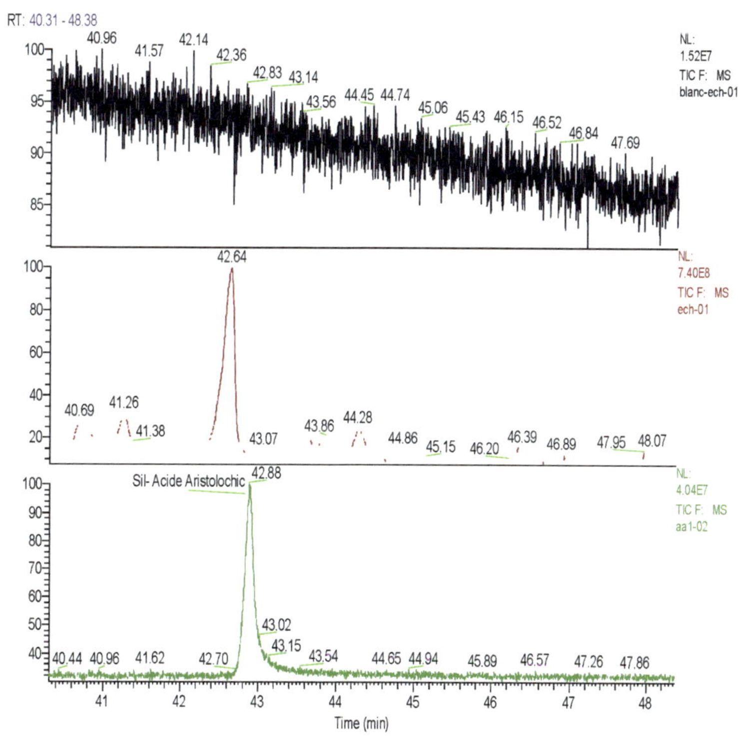

Figure 3.
CG/MS spectrum analysis of our herb (rhizome) and (AAI) standard. (Sigma A5512–25 mg Yellow powder lot # wxbb6331V^{PCODE})

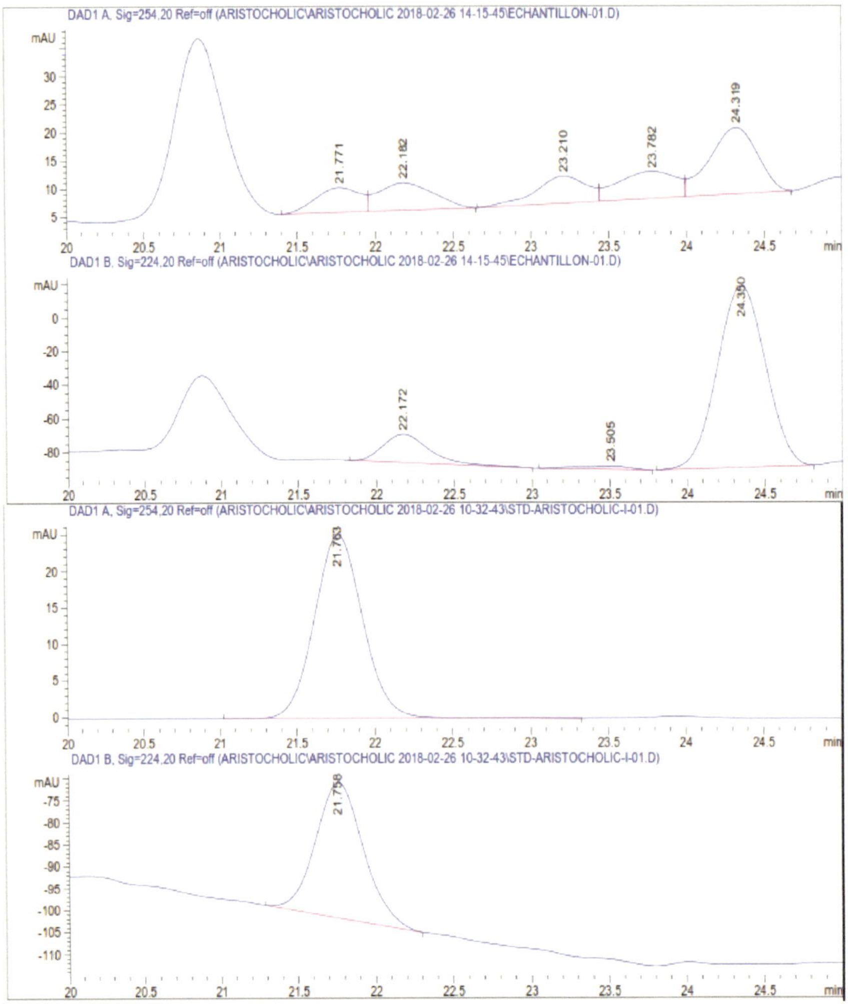

Figure 4.
HPLC spectrum analysis of bereztem roots and (AAI) standard. (Sigma A5512–25 mg Yellow powder lot # wxbb6331V^{PCODE}*)*

The Aristolochic acids content of plants varies depending for the same plant species, where it was grown, the time of year, and other factors, that might explain our results? The sample matrix components can influence the limits of detection for Aristolochic acids. We suggested also the insufficient amount of our plant we used 10 mg fine powder of bereztem roots. Although the protocol experimental described by Cherif H.S., and all has been reproduced, in our work, these authors identified the Aristolochic acid (AAI) in the rhizome of bereztem collected in the other city of Algeria: Blida (smell city) located about 47 km in southwest of Algiers. While our plant is collected in the region of Taourirt Aden in (kabylia). Broad range of biological activity of AAs, beneficial as well as adverse effects was reported by Kupchan and Doskotch [33]; therefore it would be interesting to determine the organic composition, it is known that the essential oils contained in the rhizome and aerial part of this herb are mainly responsible for the antimicrobial and cytotoxic effect [34]. And then instead to banned the use of herbal remedy containing acid aristolochic it will be possible to separate the useful from the toxic fractions of plant.

4. Conclusion

Seven elements have been determined in Aristiolochia Longa. L using INAA, technique usually quite for herbs analyses. This plant can be considered as source of trace elements for people who use it. However the traditional healers recommended the use it with care and always for short treatment periods [35]. It is clear that further experiences are planned to confirm or refute our results obtained by CG/MS and HPLC for the identification of AAs. In its warning, the FDA recommended that all botanical remedies known or suspected of containing Aristolochic acids be discarded. People should be largely aware of the regulated of some botanical products as dietary supplements by the FDA under the Dietary Supplement Health and Education Act (DSHEA) of 1994 (FDA 1995) [36].

Acknowledgements

Authors are grateful to the National Institute of Forensic Science and Criminology (INCC), Bouchaoui, Algeria for the analysis of our herb by CG/MS and HPLC.

Abbreviations

INAA	Instrumental Neutron activation analysis
CAM	Complementary and Alternative Medicine
NCP	Clinical Syndrome to the Chinese plants Nephropathy
NAA	Aristolochic Acids Nephropathy
AAs	Aristolochic Acids
AAI	Acide Aristolochique I
AAII	Acide Aristolochique (demethoxylates derivative)
EC	European Commission
CG/MS	Gas chromatography – Mass spectrometry
HPLC	High performance Liquid chromatography
IARC	International Agency for Research on Cancer
EMEA	European Medicines Agency
INCC	National Institute of Forensic Science and Criminology, Algeria
HpGe-Detector	High Purity Germanium Detector

Author details

Zohra Lamari[1]* and Houria Negache[2]

1 NUR Reactor Division, Research Department, Nuclear Research Center of Draria, Draria, Algeria

2 Nuclear Applications Division/Medical Applications Department/Nuclear Research Center of Algiers, Algiers, Algeria

*Address all correspondence to: zahra750@yahoo.com; zohralamari@yahoo.fr

References

[1] Cherif HS, Saidi F, Boutoumi H, Rouibi A, Chaouia C. Identification and characterization of some chemicals from *Aristolochia longa L*. Agricultura 2009 ; 3:76-82

[2] Saidi F, Cherif HS, Lazouri H, Aid K, Rouibi A, Bele C, Matea C. Determination of the lipid compounds of *Aristolochia Longa* L. from Algeria. Bull U Agr 2009; 66: 17-23

[3] Starr F, Starr K, Loope L. 2003. Plants of Hawai'i: *Aristolochia littoralis*. Hawaiian Ecosystems at Risk Project. http://www.hear.org/starr/reports/pdf/aristolochialittoralis.pdf. Last accessed: 10/19/05. (Support not reported. Authors affiliated with U.S. Geological Survey)

[4] Chinese HERBS Nephropathy: *AVariant Form in Japan*. Internal Medecine .2001;Vol.40, No.4 (April 2001).

[5] BenzakourG, BenkiraneN, AmraniM, Oudghiri M. Immunostimulatory potential of *Aristolochia longa* L. induced toxicity on liver, intestine and kidney in mice. J Toxicol Environ Health Sci 2011;3:214-22

[6] Sithranga Boopathy N, Kathiresan K. Anticancer drugs from marine flora: An overview. J Oncol 2010;2010:214186

[7] Robinson MM, Zhang X. The world medicines situation 2011, traditional medicines: Global situation, issues And challenges. Geneva: World Health Organization Press

[8] González F, Esquivel HE, Murcia GA and Pabón- Mora N. *Aristolochia pentandra* (Aristolochiaceae) in Colombia: Biogeographic implications and proposed synapomorphies between the pentandrous species of *Aristolochia* and its South American sister group. Rev Aca Colom Cien 2010;34:467-78

[9] Cherif H.S, F. Said et al. Identification et caracterisation de quelques Composés chimiques chez Aristolochia Longa.nr.3- 4, 2009 p. 71- 72

[10] Kohara A, Suzuki T, Honma M, Ohwada T, Hayashi M .2002 . Mutagenicity of aristolochic acid in the lambda /lacZ transgenic mouse (Muta TM Mouse). Mutat Res 515 (1-2): 63-72. support not reported. Authors affiliated with National Institute of Health Scioences , Japan ; Nagoya City University , Japan; University of Tokyo, Japan

[11] EMEA.2000. Position paper on the risks associated With the use of herbal products Containing Aristolochia species. European Agency for the Evaluation of the Medicinal Products. http:// www.emea.eu.int/pdfs/humen/hmpwp/002300en.pdf. Last accessed: 3/9/04

[12] IARC, 2002. Some Traditional Herbal Medicines,Some Mycotoxins, Naphtalene and Styrene , IARC Monographs on the Evaluation of Carcinogenic Risk of Chemicals to Humans. Vol.82, Lyon , France : International Agency for Research on Cancer.p.69-128

[13] PFAF. 2005.Plants for a future Database. Plants for a Future. http://www.ibiblio.org/pfaf/D|_search.html. Last accessed: 5/505

[14] Jean-pierre Cosyns. Aristolochic Acid-and ochratoxin A-DNA adducts: possible markers of Balkan endemic nephropathy and Associated urotherial tumors Medicine and Biology Vol.11,N°1.p; 1-4, 2004

[15] Vanherweghem JL,Depierreux M, Tielemans C , Abramowiez D, Dratwa M, Jadoul M, Richard C, Vandervelde D, Verbeelen D,

Vanhaelen-Fastre R, Vanhaelen M.1993. Rapidly progressive interstitial renal fibrosis in young women : association with slimming regimen including Chinese herbs . Lancet 341(8842): 387- 391.(support not reported, Authors affiliated with Université Libre de Bruxelles , Belgium, Vrje Universiteit Brussel, Belgium, Catholique de Louvain , Belgium)

[16] Kupchan SM, Doskotch RW.1962. Tumor inhibitors. I. Aristolochic acid, the active principle of Aristolochic India. Jmed Pharm Chem 91 : 657-659. (Supported by the National Cancer Institute. Authors affiliated with University of Wisconsin)

[17] Jackson I., Kofman S, Weiss A, Brodovsky H.1964, Aristolochic acid (NSC- 50413): Phase I clinical study. Cancer Chemother Rep 42:35-37. (Support not reported. Authors affiliated with Jefferson Medical College Hospital and Presbyterian. St. Luke's Hospital, USA)

[18] Mix DB, Guinaudeau H,Shamma M. 1982. The aristolochic acids and aristolactams. JNat Prod 45(6): 657- 666.(Support not reported Authors affiliated with Pennsylvania state University , PA; Universite de Limoges, France; Centre d'Etudes Pharmaceutiques, France)

[19] Kumar V , Poonam, Prasad AK , Parmar VS.2003. Naturally occuring aristolactams ,aristolochic acids and dioxoporphines and their biological activities. Nat P rod Rep 20(6): 565-583. (Supported by the Council of Scientific and Industrial Research. Authors affiliated with University of Delhi, India)

[20] Arlt VM, Stiborova M, Schmeiser HH .2002. Aristolochic acid as a probable human cancer hazard in Herbal remedies : A review. Mutagenesis, 17(4): 265-277

[21] Hashimoto K, Higuchi M, Makino B, Sakakibara I, Kubo M, Komatsu Y, Maruno M, Okada M.1999. Quantitative analysis of aristolochic acids, toxic compounds, contained in some medicinal plants. J E thnopharm 64: 185-189. (Support not reported. Authors affiliated with Tsumara and C company, Japan; Toyama Medical and Pharmaceutical University, Japan.)

[22] Kite GC, Yule MA, Leon C, Simmonds MS.2002. Detecting aristolochic acids in herbal remedies by lquid chromatography/serial mass spectrometry. Rapid Commun Mass Spectrom 16(6): 585-590 (support not reported. Authors affiliated with Royal Botanic Gardens, UK and University of Greenwich, UK.)

[23] IARC, 2002. Some Traditional Herbs Medicines , Some Mycotoxins, Naphtalene and Styrene, IARC Monographs on the Evaluation of Carcinogenic Risk of Chemicals to Humans .vol.82,Lyon, France International Agency for Research on Cancer .p.69 -128

[24] Grollman AP, Jelacovie B.2007. Role of environnemental toxins in endemic (Balkan) nephropathy. October .2006, Zagreb, Crotia. J Am Soc Nephrol 18(11): 2817-23.(Supported by NIEHS, Fogarty International Center and the Croation Ministries of Science and Health Authors affiliated with stony Brook University , NY; Zagreb University School of Medicine and University Hospital Center, Croatia)

[25] PozdzikAA,BertonA,SchmeiserHH, Missoum W, Decaestecker C, Salmon IJ, Vanherweghem JL, Nortier JL .2010; Aristolochic acid nephropathy revisited: A Place for innate and adaptive immunity? Histopathol., 56(4): 449-463.

[26] Mengs U.1987. Acute toxicity of aristolochic acid in redents. Aech Toxicol59 (5): 328-333(Support not

reported. Authors affiliated with Dr. Madaus Gmbh & Co., Germany)

[27] U.S.National library of Medicine – Bethesda MD August7, 2013

[28] Mengs U, Stotzem CD .1993. Renal toxicity of aristolochic acid in rats as an example of nephrotoxicity testing in routine toxicology. Arch. Toxicol., 67: 307- 311.

[29] Chung-Hsin Chen et al . Aristolochic acid – associated urotherial cancer in Taiwan PNAS /may 221,2012 /vol 109 / n° 21 /8241- 8246

[30] U.S. Department of Health and Human Services Public Health Services National Toxicology Program Research Triangle Park, NC 27709: final Report on Carcinogens Background Document for Aristolochic Acids September 2, 2008

[31] Flurer RA,Jones MB, Vela N, Ciolino LA, Wolnick KA.2001. Determination of Aristolochic Acid in Traditional Chines Medicines and Dietary Supplements , DFS/ORO/ ORA N_O.4212. U.S. Food and Drug Administration 13 pp.

[32] Tanaka A,Nishida R, Yoshida T, Koshikawa M, Goto M, Kuwahara T.2001. Outbreak of Chinese Herb nephropathy in Japan: are thre any differences from Belgium ? Intern Med 40(4): 296- 300.(Support not reported . Authors affiliated With Osaka Saiseikai Nakatsu Hospital, Japan; Kyoto University, Japan.)

[33] Kupchan SM, Doskotch RW.1962. Tumor inhibitors. I. Aristolochic acid, the active principle of Aristolochia indica. JMed Pharm Chem91: 657-659. (Supported by the National Cancer Institute. Authors affiliated with University of Wisconsin.)

[34] Jian Qing Yu, Zhi Xiong Liao et al. Composition, Antimicrobial activity and cytotoxicity of essential oils From Aristolochia mollissima. Environmental Toxicology and Pharmacology 23.2007; p. 162- 167

[35] Benchaabane A, Abbad A .1997. Medicinal Plants marked in Marrakech. Trace of the Present. Marrakech, p. 74

[36] FDA.1995. Dietary Supplement Health and Education Act of 1994. U.S.Food and Drug Administration. http://www.cfsan.fda.gov/dms/dietsupp.html Last accessed: 7/1/06